The Modern Amateur Astronomer

Patrick Moore (Ed.)

Springer

Cover photograph by courtesy of SCS Astro of Wellington, Somerset, England

ISBN 3-540-19900-4 Springer-Verlag Berlin Heidelberg New York
ISBN 0-387-19900-4 Springer-Verlag New York Berlin Heidelberg

British Library Cataloguing in Publication Data
Modern Amateur Astronomer. – (Practical
Astronomy Series)
 I. Moore, Patrick II. Series
 522
ISBN 3-540-19900-4

Library of Congress Cataloging-in-Publication Data
The modern amateur astronomer / Patrick Moore (ed.).
        p.   cm. -- (Practical astronomy)
 Includes index.
 ISBN 3-540-19900-4 (pbk. : alk. paper)
 1. Astronomical instruments.   2. Spherical Astronomy.
3. Telescopes.   I. Moore, Patrick.   II. Series.
QB86.M63   1995   95-32987
522--dc20   CIP

© Springer-Verlag London Limited 1995
Printed in Great Britain

Typeset by Editburo, Lewes, East Sussex, England
Printed by the Alden Press, Osney Mead, Oxford
34/3830-543210 Printed on acid-free paper

Contents

Introduction

Astronomy has always been one of the few sciences in which the amateur can make valuable contributions. Indeed, not so very long ago some of the world's leading astronomers were amateurs – the classic example being the third Earl of Rosse who, alone and unaided, built what was then the largest telescope in the world, and from 1845 used it to make fundamental discoveries.

This is still true today, if to a somewhat more limited extent. There are of course amateur theorists and cosmologists, but all in all the main amateur contribution at the present time comes from the observers. It is quite correct to say that the average amateur knows the night sky a great deal better than the average professional, who spends his observing time looking at a television screen, and who very seldom puts his eye to an eyepiece!

Yet there has been a tremendous change in quite recent times. Forty years ago, the average amateur astronomer was equipped with a modest telescope, and carried out almost all of his work visually. Invaluable observations were made of Solar System bodies, and of variable stars; some amateurs made their reputations by hunting for comets and novæ, with remarkable success. Now, much of this has altered. The telescope remains the essential tool, but the modern amateur can make use of techniques which would have been completely beyond his reach (or his personal capability) a few decades ago – and he can use modern equipment which did not even exist in the 'pre-electronic' age. Sophisticated photography has come into use, plus electronic devices; for example, there are many amateurs who own CCDs (charge-coupled devices, which form the heart of a very sensitive television camera – more about this in Chapter 7) and are fully able to produce work of professional standard.

There are many books which introduce the subject of practical astronomy. There are also many which cater for a far higher technical level. However, there is a gulf between these two types, and I hope that this gulf will be filled by the two present books – this one and its companion, *The Observational Amateur Astronomer*. In both books every chapter has been written by an author who is really experienced in his chosen field.

Each chapter has been written from a personal perspective, and in editing this book I have been careful to leave the writers' ideas, preferences, and writing styles alone. The same topics may be touched upon by more than one author, often from a slightly different perspective. This is quite deliberate.

I expect that readers of this book will have some prior knowledge of amateur astronomy, but there is nothing here which will puzzle any enthusiast who has taken the trouble to master the basic facts.

Remember, too, that amateur work is warmly welcomed by the professionals. Amateurs carry out research which the professional does not want to do, has no time to do, or literally cannot do.

There is always something new to find out.

Patrick Moore, 1995

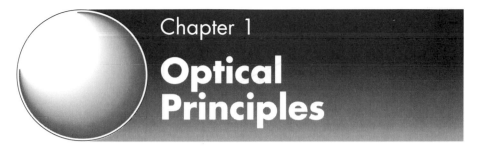

Chapter 1

Optical Principles

R. W. Forrest

Why learn about the optics of a telescope? The answer is simple: an appreciation of some of the optical principles that underlie astronomical telescopes will help serious observers make the best use of an instrument, and make reasoned judgements as to its capabilities for fulfilling particular observing aims. An understanding of how a telescope (or other optical instrument) works is central to an understanding of what it can do.

Optical Theory

Let's begin with the meanings of some fundamental terms. Telescopes using lenses as their principal components are commonly called *refractors*. Those utilising mirrors are *reflectors*. When both lenses (other than an eyepiece) *and* mirrors are essential to the performance the telescope is termed *catadioptric*. The lens of a refractor is called its *objective* or *object glass* (often abbreviated OG). The first mirror of a reflector is the *primary* and subsequent mirrors are termed *secondary*, and then *tertiary*, etc.

The *aperture* of a telescope is the diameter of the light beam that it collects, and is usually the diameter of its objective or primary mirror. Bigger telescopes allow fainter stars to be studied and show finer detail, which is why the aperture is so important that references to a telescope always include it – e.g., a '200-mm Schmidt–Cassegrain' means a telescope of aperture 200-mm.

The function of any optical system is to gather

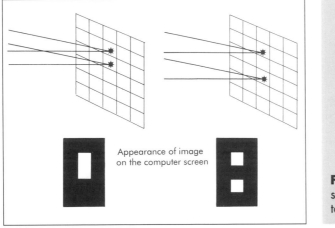

Appearance of image on the computer screen

Figure 1.3 Two stars imaged close together on a CCD.

Telescopes with the smallest values are described as *fast*, because their brighter image permits photographic exposures to be made more quickly.

Quality of Imaging

The ideal that the image can be made up of infinitely small points arranged in an undistorted pattern can never actually be realised in practice. Errors arise from several sources: the wave nature of light, the geometry of the optical surfaces, the accuracy with which that geometry is realised during the grinding of the surfaces, the accuracy with which the surfaces are aligned with one another and, for refractive designs, dispersion. Finally the passage of the light through the atmosphere creates severe problems even before it reaches the telescope optics.

The Diffraction Limit

The wave nature of light sets a fundamental limit on the size of the points that make up the image. *Diffraction* – the phenomenon of light spreading a little beyond the limits of what geometry leads us to expect – at the aperture of the telescope causes the light originating from one direction to be imaged into a region of finite size. If the aperture is a simple circle (e.g. a refractor) then each image 'point' will in fact be a tiny bright disc, known as the Airy disc, surrounded by a set of fainter rings (Figure 1.5).

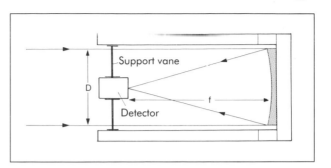

Figure 1.4
Schematic of a
mirror used at
prime focus.

The brightness of the image is greatest in the centre of the Airy disc and then falls to zero at its boundary before rising again to form the first ring. It is clear that the Airy discs of two adjacent points of the object will overlap and that the amount of this overlap will determine the observer's ability to see the points as distinct from one another. The resolution of the telescope is taken as the angular separation of two points where the centre of the Airy disc for one of them coincides with the zero intensity point between the disc and first ring of the other (the Rayleigh criterion). This occurs for an angular separation a (measured in radians) given by

$$a = 1.22\lambda/D,$$

where λ is the wavelength of the light. (Note that D and λ must be expressed in the same units and that 'light' here means any electromagnetic radiation so that the formula applies also to radio telescopes.) This angle is smaller when D is bigger and hence bigger telescopes have a better resolution. Because D is always much larger than λ for optical telescopes, the value of the angle is very small. In this case the more useful expression is

$$a = 251\ 570\lambda/D,$$

giving a in arc seconds, and this approximates to the rule of thumb that the resolution is 1" for 100 mm of aperture, 0.5" for 200 mm of aperture, etc., at a wavelength of 390 nm (0.39 mm or 0.000 39 m). This wavelength lies in the violet; the performance of the instrument will be worse in the red – 1.6" for 100 mm of aperture at 650 nm – other things being equal. It should be realised that the Rayleigh criterion gives a convenient theoretical value for the resolution, but that in practice the relative brightnesses of the adjacent image points are also important. Thus, for example, a small, bright spot on a planetary surface might be detectable while a similar-sized spot little different in brightness within half of the wavelength of light and mirror surfaces to within an eighth of the wavelength. By 'correct', he meant that nowhere would the surface depart from the perfect curve by more than the given amount. Experiments with small reflectors have shown that in all but superlative seeing conditions this accuracy is not necessary, and a mirror with overall deviations of not more that a quarter of a wavelength will perform creditably. Further, a more useful criterion of quality than the overall error is the *r.m.s. deviation*, which takes account of the amount of surface affected. It is intuitively obvious that a surface with a single error over one small area should perform better than one which has ripples all over. Recast in these terms a satisfactory performance will be achieved if the r.m.s. surface deviation is no more than about '$^1/_{13}$ wave' (i.e., $^1/_{13}$ of the wavelength of light) for a mirror or $^1/_4$ wave for a lens.

Alignment

If the telescope optics are to perform to the limits permitted by their accuracy, it is important that all the components are carefully aligned with one another. Failure to achieve and maintain optical alignment (optical alignment is often called 'collimation') will introduce spurious aberrations, coma normally being the most obvious. Almost equally important – for large telescopes at least – is the requirement that there should not be any mechanical stress put on the components by the mounting; the resulting strain can distort the lens or mirror enough to affect the image.

For amateurs who have bought reasonable telescopes from a reputable manufacturer (see Patrick Moore's comments in Chapter 2!) I should stress that, with the exception of fast paraboloids (coma) or wide-field eyepieces (astigmatism), optical aberrations should not be noticeable and *incorrect collimation or mounting should be suspected if they appear.*

The Atmosphere

All that I have said so far assumes that a perfect plane wave is incident on the telescope aperture. Unhappily – and with the spectacular exception of the Hubble Space Telescope – this condition is only fleetingly achieved because passage of light through the turbulent atmosphere corrugates the wavefront in a random and quickly changing fashion.

The effect that this has on the image depends on the size of the telescope and the prevailing conditions but includes both image movement and defocusing. In all but the most stable conditions, however, the average distribution of light is in a *seeing disc* much larger than the Airy disc. It is for this reason that the resolution attainable in photographs is limited (and more or less predictable), whereas the skilled visual observer can combine the information presented during the brief instants when the image is steady into a detailed drawing.

Professional observatories are beginning to solve the problem by the use of adaptive optics, in which key elements of the optical systems are continuously monitored and balanced by a computer to give an optimum image all the time. A simpler device acting as an image stabiliser is available for amateur use on the planets or brightest stars.

Vignetting

In a properly designed system every point on the detector (or eyepiece field) should be able to receive light rays from the whole of the aperture. If this condition is not fulfilled then objects towards the edge of the field will be fainter. For example, a camera adapter might have too small a central aperture or a secondary mirror be too small in diameter. While this effect can be removed for cosmetic purposes when a CCD is being used, by the process of *flat-fielding*, the affected area will have a poorer signal-to-noise ratio (Figure 1.8).

Image Contrast

The diffraction rings surrounding the Airy disc slightly reduce the contrast in extended images (an extended image is one that is resolved other than as a stellar point, for example nebulæ and planets). If the telescope is one that has a partial obstruction to its light path (most reflectors) then the disc will be slightly less bright and the rings brighter, with a further reduction in image contrast. The performance then becomes equivalent to that of a telescope whose diameter is the same as the actual diameter, minus the obstruction diameter. If the secondary of a reflector is supported by a *spider* con-

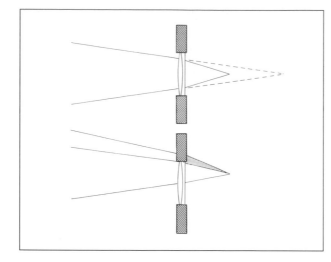

Figure 1.8 An example of vignetting. The telecompressor lens is too small to transmit all the rays for the off-axis point shown in the lower drawing and the shaded portion of the beam does not reach the focal plane.

sisting of three or four supporting blades or struts, then the characteristic diffraction 'spikes' you will have noticed in photographs are produced around bright stars. While the effect on image contrast is small, suitable design of the spider can minimise it, and both imperfections can be eliminated by use of an off-axis construction.

Of more serious concern for faint objects is the effect of direct skylight reaching the detector (or eyepiece). This is difficult to avoid in Cassegrain types and is exacerbated if the sky is bright because of *light pollution* (spilled light from cities, motorway lighting, etc.). The solution is to use baffles to prevent the skylight from reaching the focal plane, but care is required not to introduce vignetting, and light can still glancingly reflect off the sides of the baffle tube, even if it is painted matt black.

Types of Telescope

There are many different designs of telescope, but I will concentrate on the telescope configurations most used by amateurs.

Refractors

The refractor, the instrument most people think of when you say the word 'telescope', consists of a rigid tube with the objective at one end and the eyepiece at the other, and has much to commend it. It gives high-contrast images and is probably the optimum instrument for solar system studies. There are two big disadvantages: the length of the tube and the price.

Reflectors

The *Newtonian* reflector uses a parabolic primary mirror, together with a flat secondary mirror to divert the light to the side of the tube (Figure 1.9). This means that the detector (or eyepiece) does not face directly up at the sky, giving good contrast. The Newtonian is very versatile and inexpensive. Deep-sky photographers will choose a fast mirror (f/6, perhaps), which allows a shorter instrument that may be sufficiently portable to transport to a dark site, while planetary observers will choose a longer focal ratio (f/10), which allows a smaller secondary mirror to be used. There aren't many disadvantages to the Newtonian: one of the few is the awkwardness of attaching auxiliary equipment at the focus.

In the *Cassegrain* design a parabolic primary is again used, but this time the secondary is a convex hyperboloid placed inside the prime focus (Figure 1.10). The final image is produced behind the primary, which has a hole drilled through the middle to allow the light through. The shape of the secondary is chosen to give a perfect image on-axis at a specified distance behind the primary. The design has been widely used professionally for the convenience it offers for attaching instrumentation, and for the very short tube relative to the focal length, a consequence of the fact that the hyperboloid secondary increases the effective focal length. The distance that the secondary is inside the prime focus, divided into the distance between the secondary and the actual focal point behind the primary, gives the *amplifi-*

Figure 1.9 ∧ 305-mm f/6.6 Newtonian telescope. The primary mirror is within the boxed-in section of the tube at the left; a small part of it is brightly illuminated. The elliptical secondary set at 45° to the axis is clearly seen towards the right. It reflects the converging beam from the primary towards the eyepiece, which is seen at the extreme right.

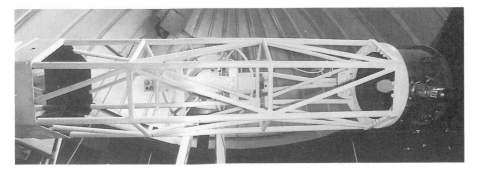

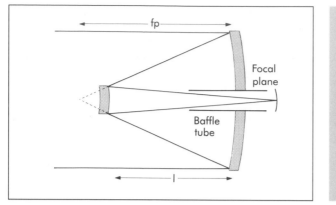

Figure 1.10 The arrangement of a Cassegrain telescope. The Cassegrain focal length f is given by $f = f_s f_p/(f_s - f_p + l)$, l being the intermirror distance.

cation factor, typically between 3 and 6. The effective focal length of a Cassegrain telescope is the focal length of the primary multiplied by the amplification factor. Another convenience is that the focus position is readily adjustable over quite a wide range by changing the distance between the mirrors (generally by moving the secondary).

The development of *Schmidt–Cassegrain* catadioptric telescopes made the advantages of the Cassegrain widely available to amateurs. In these, a *corrector plate* is placed inside the prime focus of the spherical primary and supports the secondary. The fast ($f/2$) primary means that an exceptionally compact and hence portable instrument is produced, and the spherical primary is cheaper to make than the paraboloid required by the Newtonian. The primary is moved for focusing in these instruments and the focal plane movement δp is

$$\delta p = d \times m^2,$$

where δd is the movement of the primary (typically 0.8 to 0.9 mm per turn of the focus knob), and m is the secondary magnification (amplification), typically 5. The movement varies m itself and hence the effective focal length can also change significantly. The focal length change δf produced by moving the focus position by δp is given by

$$\delta f = \delta p \times f_p/f_s,$$

where f_p and f_s are the focal lengths of the primary and secondary respectively. For a 200 mm $f/10$ instrument, f_p/f_s is 2.3. This means that moving the focus out by 60 mm (which might be necessary when using a telecompressor) increases the focal length by 10%. The focal plane is slightly curved, but combined telecompressors and flatteners are available to combat this. As noted above, direct skylight can be troublesome, but the latest models incorporate extra baffles to reduce this.

In the Schmidt–Cassegrain design the corrector plate has an aspherical shape that is difficult to manufacture and test. A development of this design, the *Schmidt–Maksutov*, uses a corrector plate that is a deeply curved meniscus lens, mounted with the concave side facing away from the primary. The secondary is simply an aluminised spot in the centre of the corrector plate, and the resulting telescope is much easier (and more important, much cheaper) to manufacture.

Eyepieces

If the telescope is to be used for visual observations, or if close-up photography is required, then an eyepiece will be necessary. An eyepiece collects all the light from the focal plane image and converts it back to parallel beams suitable for the eye to focus. The beams all pass through a small area just beyond the eye lens – the exit pupil – and this is where the eye should be placed for optimum views (Figure 1.11). The distance behind the last lens surface and the exit pupil is called the eye relief. Spectacle wearers will find eyepieces of greatest eye relief the most comfortable to use, with least danger of bumping into the lens. Eyepieces offering long eye relief will have the largest eye lenses.

The magnification produced is the increase in the apparent angle between two objects. It is easily calculated from

$$\text{magnification} = f/f_e$$

where f_e is the eyepiece focal length (which will be engraved upon it). (Remember, however, that f for a Cassegrain-type telescope will change as the focus position is varied to accommodate different eyepieces.) To extract the maximum information from the image

Figure 1.11
Emergent beams
from the eyepiece.

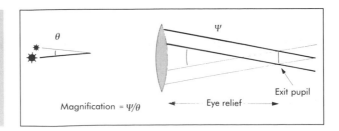

θ

ψ

Exit pupil

Magnification = ψ/θ

Eye relief

(when the atmospheric seeing is very good), the magnification must be large enough for the eye to see detail on the scale of the Airy discs produced by the telescope. In practice this implies a magnification numerically equal to the aperture of the telescope in millimetres, or greater. For example, a 200-mm telescope requires at least 200×. If this were an $f/10$ system its focal length would be 2000 mm, so the eyepiece focal length needed would be $f_e = f/$magnification $= 2000/200 = 10$ mm, or less.

An important practical factor about an eyepiece is its angular field of view – the amount of sky visible through it. In general this will be smaller for higher magnifications, and it will also tend to be smaller for eyepieces mounted in a smaller barrel. For use with telescopes with focal lengths above 2 m a 2-inch (50-mm) diameter style is to be preferred, at least for low magnification. Knowing the field of view (f.o.v.) will assist in correlating the telescopic view with finding charts, so it is worth measuring it for each eyepiece. This may readily be done by timing how long it takes a star of known declination δ to cross the field centrally. Then f.o.v. measured in arc minutes is given by:

$$\text{f.o.v.} = t/(4 \cos \delta)$$

when t is expressed in seconds.

There is a wide selection of eyepieces available, ranging from inexpensive *Kellners* suitable for low power, through *orthoscopics* giving excellent higher power performance, and the *Plössl* and *Nagler* types that offer superb wide-angle views. The latter two types tend to be expensive, though. For work such as planetary sketching or close double stars it is worth experimenting with just a single simple lens as an eyepiece. The field of good definition will be limited, but this is of no consequence, and it may well be found to give the best results over the limited area. The fewer surfaces there are, the less chance there is for loss of light and the formation of ghosts, though these are minimised in all eyepieces today by careful design and the use of anti-reflective coatings.

Other Equipment

For experimentation purposes, any equipment that has a lens built in may be placed where the eye would normally go. Set the lens to infinity first and adjust the telescope focus for the sharpest view.

Chapter 2
Buying a Telescope

Patrick Moore

It is fair to say that, sooner or later, everyone seriously interested in astronomy will want to buy a telescope – or, alternatively, make one. This is often where the trouble begins, because commercially produced telescopes are either good or cheap: not both. And there are various pitfalls which have to be considered.

In my view (and I am open to challenge here), the smallest telescopes which are of any real astronomical use have apertures of 3 inches (7.5 cm) for refractors, and 6 inches (15 cm) for Newtonian reflectors. Smaller instruments simply do not collect enough light, quite apart from their numerous other disadvantages – such as shaky mountings, which are usually about as stable as blancmanges. All in all, the cost of a useful telescope (at the time of writing) is bound to be at least £350 – or in the USA about $500 – and this sounds a great deal; but it must be remembered that the cost is non-recurring, since if well treated a telescope will last for a lifetime. Moreover, the cost does not sound so great when you compare it with, say, the price of a couple of rail tickets between London and Edinburgh!

Much the best procedure, if finance is a real problem, is to begin with binoculars, which have most of the advantages of very small telescopes but few of the drawbacks. Binoculars are classified according to their magnifying power and the diameter of the main objectives. The magnification is fixed (except in zoom binoculars, which on the whole are not to be recommended astronomically), and so at the outset the buyer must be very sure of just what he or she wants. The diameter of the objective is always given in millimetres. Thus a pair of

7 × 50 binoculars means that the magnification is 7 times, and each lens is 50 mm across, and so on.

It might be thought that the ideal is to have a really high magnification, but this is not so. A high power means a relatively large aperture, so that binoculars become heavy, and some sort of mounting is essential. With, say, a 20 × 80 pair you will need a tripod stand, though with lower powers a neck mounting is adequate. All in all, I would recommend a mounting for any binoculars of magnification over ×12.

Next, the field of view: if you want extended star-fields, then select a magnification of ×10 or less. These pairs are suitable for terrestrial use. If you are going to buy only one pair, I would say that 7 × 50 is about right. Second-hand binoculars can be bought, and have the advantage that they can be tested 'on the spot'; check to make sure that both lenses can be brought to sharp focus – one lens is usually adjustable – and if you cannot obtain a sharp image with both eyes simultaneously or if you have difficulty 'merging' the images from the two sides of the binoculars, then abandon them at once. Make sure, too, that there is no false colour, and that the lenses are firm, not loose in the frame.

Binoculars are much more useful astronomically than is generally believed; they can be suitable for some variable star work, for example. Their main drawback is sheer lack of magnification, so that they are unsuited for planetary observation (though they will give splendid views of the Moon).

The thing *not* to do is to go to a camera shop or general store and pay a considerable sum – perhaps up to £200 – for a nice-looking, well-finished telescope with an aperture of around $2^1/_2$ inches (60 mm) for a refractor or 4 inches (100 mm) for a Newtonian reflector. Generally, these telescopes will be difficult to set up, and even when this has been achieved they will be unsteady; the light-grasp is bound to be poor, and the field of view small. In particular, beware of misleading advertisements. One often finds a telescope advertised as 'up to 300 times magnification', for instance. If nothing is said about the aperture, have nothing to do with it. A good rule is that no telescope can be used effectively with a magnification of more than ×50 per inch of aperture, so the most that a 2-inch can ever bear is ×100 – and then only if the optics are first-class.

Unfortunately, good second-hand telescopes are now about as common as great auks. Never buy one without having it checked – and remember, a poor telescope does not immediately betray itself by its outward appearance (particularly in the case of a reflector). The next alternative is to make your own telescope. Frankly, lens-grinding is a task for the real expert, and to buy an object-glass and mount it is not really sensible, if only because the main cost of a refractor is in the object-glass itself. On the other hand, making a mirror for a Newtonian is much less of a problem, and is more laborious than difficult – see Chapter 5 in this book! And if you buy the optics for a Newtonian, mounting them is not a problem for anyone who (unlike myself) is reasonably good with his hands.

Everything depends upon the kind of observing you want to do. Portability is one problem. If you live in a city, the only object which is really available is the Sun; otherwise you will have to take the telescope away from inconvenient lights, so you need something portable. A 3-inch refractor on a tripod is manageable, and so is a 6-inch Newtonian, which can almost certainly be dismantled for transport; but for anything much larger than this the difficulties are more acute, unless the Dobsonian pattern is followed.

For the observer who intends to do no more than take the occasional casual 'look around the sky', a 3-inch refractor is ideal. On the other hand, it will not stand a magnification of more than ×150, so that for planetary work it is limited. However, it is very suitable for observations such as grazing occultations.

A 3-inch Newtonian is of little use; if you want a minimum-size reflector, then 6 inches is the lower limit. This costs about the same as a 3-inch refractor, and it may well be that the would-be purchaser has a direct choice between the two. If he or she is concerned with observing the Sun (by projection of course), then the refractor wins hands-down; for deep-sky observations the reflector is far better.

The simple altazimuth mounting (more detail later, in Chapter 3) has its advantages for a small telescope; the Dobsonian is a variant of this. There is not a great deal to go wrong, and the telescope can be swung quickly from one part of the sky to another with any complicated manœuvres; neither is there any need to spend a long time in adjusting the polar axis (there is nothing more infuriating than a badly aligned equatorial). Against this, the telescope must be moved constantly in two directions to allow for diurnal motion, which is obviously awkward.

Many of the small telescopes which I so despise are

mounted upon attractive-looking equatorials. Alas, they are generally so shaky and difficult to align that they are more trouble than they are worth. But a large telescope should be equatorial and driven – and if not, then there is no real point in trying to use it for photography.

So let me try to sum up my recommendations.

- Decide how interested you really are. Begin by investing in a pair of binoculars, which do not cost much, and 'look around'.
- Avoid the temptation to buy a small, nice-looking telescope from a camera shop or general store. Do not pay much for any refractor below 3 inches in aperture or any Newtonian below 6 inches. Of course, a smaller telescope is better than nothing at all, but I would vastly prefer binoculars.
- If you want to buy a telescope, go to a recognised manufacturer, and check first; it is sad but true that some commercially produced telescopes are of poor quality – particularly reflectors. And remember, a bad telescope is not always readily identifiable.
- Decide what you really want. If you are a solar enthusiast, go for a refractor. If deep-sky is your main interest, then a reflector is to be preferred. For planets – well, I am very refractor-minded, but you need at least 4 inches of aperture, which is expensive and more or less non-portable.
- If finance is not a problem, then something in the nature of a Schmidt–Cassegrain or a Maksutov is ideal. Again, go to a reputable supplier.

Finally, weigh everything up carefully. A telescope is something which will last you a lifetime, so make absolutely sure what you want before buying!

Chapter 3

Commercially Made Telescopes

John Watson

First Principles

You're setting off to buy a telescope – these are the things you need to think about:

Aperture

Remember that the larger the aperture of a telescope, the more light can get in, the brighter the image will appear, and the fainter the objects you will be able to see. Equally important for astronomers is the fact that the *resolving power* (the amount of detail you can see) is also limited by the aperture. Follow Patrick Moore's guidelines about aperture (see Chapter 2), and go for a minimum of 150 mm (6 inches) for a reflector, and 75 mm (3 inches) for a refractor.

Quality

Not all lenses or mirrors are created equal. Bad optics of *any* size will give poor images. Mostly, you get what you pay for.

Mounting

There is nothing worse than a telescope you can't point at the object you want to observe. A mounting that is rough or jerky, or wobbles, will always be an unending source of irritation and frustration.

There are two main kinds of mounting, the *altazimuth* and the *equatorial* mounting.

The altazimuth allows the telescope to swing in a horizontal and in a vertical plane. Access to stars immediately overhead is often a problem with refractors that have altazimuth mountings.

The mounting may be fitted with *slow-motion* gearing, usually operated by knobs, which enable you to move the telescope very slowly up and down or left and right. Slow-motion controls are very useful, but need to work smoothly and without backlash.

The limitations of an altazimuth mounting become obvious if you want to observe an object for a long time or under high magnification: you have to manipulate *both* knobs to follow objects as they move across the sky, a feat of coordination that I at least find very difficult.

Most commercially produced altazimuth-mounted telescopes use a *fork mounting*, in which the telescope tube is supported on either side by a solid fork.

A rather special kind of altazimuth mounting has become popular for large reflectors. This is the *Dobsonian* mounting. It was invented by an American monk called Dobson, for use with large, simple reflectors. The Dobsonian mounting relies on super-smooth bearings made out of a plastic called polytetrafluoroethylene (PTFE, sometimes known by the brand name Teflon), which is hard and very slippery. The reflector must be accurately balanced to stay poised at any angle but, with this done, even a large, heavy telescope can be moved in any direction – very delicately if necessary – with your fingertips. A typical Dobsonian is shown in Figure 3.1.

Dobsonian mountings have four great virtues: they are cheap, they are compact, and they are very quick and easy to set up. You don't see many for sale, but they are easy to make yourself if you have even the most basic carpentry skills. Look out for 'do-it-yourself' projects, designs to copy, or even kits in the astronomical magazines. If your interest is in 'sky-gazing', or in looking at deep-space objects such as nebulæ and galaxies, then a Dobsonian mounting might well be a good choice.

An *equatorial* mounting is essentially an altazimuth mounting tipped over at an angle so that the once-horizontal axis is lined up with the axis of rotation of the earth: this axis is known as the *polar axis*. Stars and

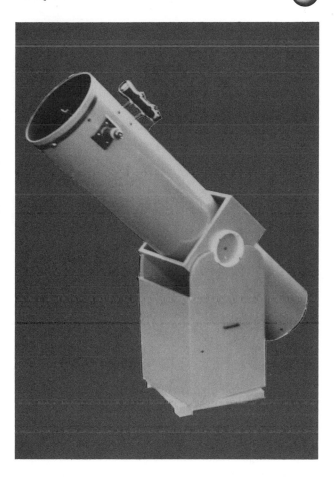

Figure 3.1
A Dobsonian
telescope.

planets can then be tracked accurately by rotation about this axis only, which makes it possible to hold an object stationary in the telescope's field of view indefinitely.

If the polar axis is turned by a small motor, then star tracking (approximately, at least) can be made automatic.

It isn't quite as simple as it seems to design an equatorial mounting, because the telescope has to *balance* in every direction so that it can be turned smoothly. An equatorial mounting is essential for any form of astronomical photography, and a great asset for almost any other kind of serious observing. Commercially produced equatorial mountings are usually either *fork mountings* – Figure 3.2 – or *German (Fraunhofer) mountings* – Figure 3.3.

Fork mountings are compact and are ideal for Cassegrain telescopes, or for large Newtonians. The German (or Fraunhofer) mounting uses a counterweight to balance the mass of the telescope, and so is

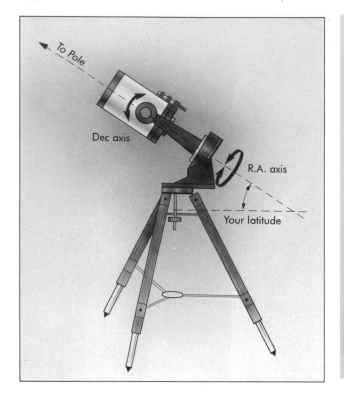

To Pole

Dec axis

R.A. axis

Your latitude

Figure 3.2
A typical fork
mounting.

both heavier and larger than the fork. However, it allows easy viewing of all parts of the sky and is the only sensible equatorial mounting for small refractors.

Dobsonians (and some other altazimuths) can be 'converted' to track stars like an equatorial mounting by standing them on an *equatorial platform*, a low (just a few inches) platform that tilts in such a way that it simulates the motion of an equatorial mounting for up to about half an hour before it needs to be reset.

Equatorial platforms can be purchased from a number of suppliers, but are quite expensive.

Convenience

You should think about how convenient your telescope is for you to use. If you are not going to use high magnifications or take photographs then the time taken to set up a heavy equatorial mounting may not be worthwhile. A more portable instrument might be better for you. If you are going to spend your time looking at large faint objects a Dobsonian might be what you need. On the other hand, if your interest is in astronomical pho-

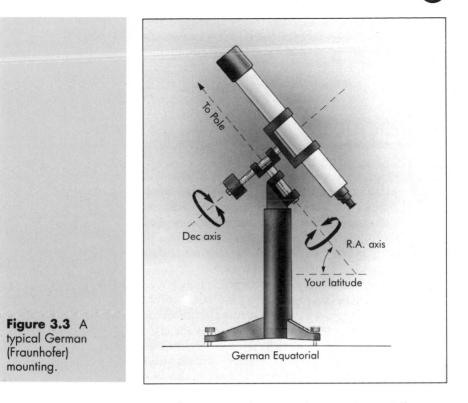

Figure 3.3 A typical German (Fraunhofer) mounting.

tography, you might seriously consider building (or having built) some kind of observatory or shelter to protect a more permanent installation.

If you live in a city or town, you will probably have to lug your telescope to an observing site before you can use it. Bear in mind that big telescopes are *heavy*. An averagely fit man wouldn't want to carry even a lightly-constructed 300-mm (12-inch) reflector more than a few yards on his own.

Magnification

Magnification is *not* important in specifying a telescope, but newcomers to astronomy often think it is, which is why I am mentioning it. The *magnification* of a telescope is determined *only* by the focal length of the objective divided by the focal length of the eyepiece. Thus an objective lens (or primary mirror in a reflector) of 1 m focal length, fitted with an eyepiece of 25 mm focal length, would give a magnification of 1000/25 = 40. A magnification of 40× is quite sensible for observing stars through a 150-mm (6-inch) telescope.

The *maximum* magnification that is usable is determined *only* by the diameter of the objective lens or primary mirror (the aperture). Again, remember Patrick Moore's guideline – *the highest usable magnification is about 50× per inch of aperture.*

Binoculars

For many kinds of astronomy, binoculars are a better first investment than a telescope. If you are interested in comet hunting, for example, binoculars are far more use to you than most telescopes. If you are a 'casual' astronomer and want to absorb the beauty of the night sky and learn to find your way round the constellations, then binoculars are a good start and excellent value for money.

For my money, don't go for high magnification, but *do* go for aperture. A magnification of ×10 is about the highest that it is sensible to buy. Higher magnifications are unlikely to give you a better view of anything in the night sky, but make it very difficult to hold the binoculars still. An aperture of 50 mm or more is perfect – you shouldn't buy anything less then 35 mm for astronomical use.

Zoom (variable-magnification) binoculars are available, but in general should be avoided for astronomy. They incorporate variable magnification eyepieces which – cost for cost – are more complex and give a worse performance than simple eyepieces. Zoom doesn't add much – through binoculars, a star (or even a planet) looks much the same under 'high' magnification as it does under low magnification!

Field of view is important. Compare different binoculars to see how large a field of view they have.

Remember that important factor that is seldom mentioned by the salesman: the *eye clearance*, or how far you can hold the eyepieces away from your eyes and still see the full field of view. Always check this before buying binoculars, especially if you wear glasses.

Finally – and it follows from what I have said above – don't buy binoculars by mail order unless you are quite sure of what you are getting.

Small Refractors

Patrick Moore is right to dislike – no, hate is nearer – the small 'astronomical' refractors of between 37 and 75 mm (1¹/₂ to 3 inches) aperture that are sold by non specialist shops. I am sure they have put thousands of people off astronomy. They are often finished to a very high standard in coloured enamel, mounted on a varnished wooden tripod, and fitted with a small (often 20 mm or less) 'guide' telescope. Sometimes very high maximum magnifications (300× or more) are quoted.

In general, such telescopes are almost completely useless. It is difficult to test the optics in the shop, as a view of the stars is much less tolerant of poor lenses than a view of the street. My rule of thumb is to set the telescope up on its mounting and tap the front of the tube with my hand. If the telescope tube wobbles appreciably, then the mounting is inferior and it's a good bet that the optics are as well.

There are some excellent small refractors on the market (such as the Odyssey and Vixen ranges), but they are quite expensive and are usually sold only by specialist astronomical suppliers. Such instruments are often of professional quality. Once you have seen and handled one there is no chance that you will ever mix it up with the other sort!

Small Reflectors

Small Newtonian reflectors – not unlike the small refractors described above, but with apertures in the range of 75 to 100 mm (3 to 4 inches) – are also quite common in non-specialist shops and tend, in general, to be equally poor performers.

Again, very workmanlike small Newtonian telescopes are also sold, but most often by specialist astronomical suppliers. Paradoxically, some of the better ones (but not the best) are often relatively crude in terms of finish and presentation, having been made by small local companies rather than by factories in the Far East. A solid, shake-free mounting is always a good sign.

Catadioptric Telescopes

Catadioptric telescopes – instruments that combine a

primary mirror with a full-aperture lens or corrector, remember – have many of the advantages of reflectors and refractors, and are the nearest thing to 'the best of both worlds' for amateurs. They are also shorter by up to 50% than the equivalent Newtonian reflector, which makes them convenient to store and to move around.

Catadioptric telescopes are more complicated, and thus more expensive, than Newtonians of equivalent aperture, but they are cheaper – and in the larger sizes much cheaper – than an equivalent refractor. Apertures vary from about 75 mm (3 inches) to 375 mm (15 inches) and more.

Equatorial mountings are the rule, or balanced altazimuth mountings that can be tipped over to make an equatorial mounting with the aid of an 'equatorial wedge', sometimes supplied as an accessory. Most have optional motor drive on the polar axis, most have optional motor-driven slow motions on both axes, and many have optional computer control (see Chapter 7).

Larger Telescopes

Commercially made large telescopes are expensive. It is cheaper to make one yourself (see Chapter 5), but you will need a certain amount of skill and lots of time.

The best place to start is with the astronomical magazines – *Sky & Telescope*, *Astronomy*, *Astronomy Now*, *Practical Astronomy* and all the others will have many advertisements and occasional product reviews.

Don't buy an instrument like this lightly (even if you are rich!); read as much as you can before you make your purchase, and if you are in any way uncertain, take advice from your local astronomical society first.

To Summarise ...

It is never too early to ask yourself what you want to look at through your telescope. Different instruments are suitable for different applications. If you are interested primarily in the Moon and the planets then you should probably try for a refractor – or at least a telescope with a long focal length and a small aperture – $f/8$ or more.

On the other hand, you may want to observe extended objects – nebulæ, star clusters, and galaxies – in which case light grasp is all-important and you should go for aperture. A Dobsonian might be ideal.

Photography of any sort requires an equatorial mounting, and a mechanical or electric drive is almost essential as well. It takes a long time to set up an equatorial mounting accurately in line with the Earth's polar axis, so you if you want to take photographs of the night sky you should seriously think about a permanent site for your telescope. The next chapter looks in more detail at the possibility of building your own observatory.

If you have no particular opinion, buy a general-purpose instrument – if such a thing exists. In my opinion, either an equatorially mounted 8-inch Newtonian or (if you have more money to spend) a Schmidt–Cassegrain comes nearest.

Chapter 4

Observatories and Telescopes

Denis Buczynski

If you can use a portable telescope, why go to the considerable effort of designing and building a permanent shelter?

Amateur astronomers have many observational interests, and consequently choose telescopes which are suited to their needs. These range from small refractors through folded catadioptric systems to reflectors large and small. Each has advantages and drawbacks, but all can be used in a variety of ways, be it visually, photographically or electronically (using CCDs or photometers). No specific type of telescope can excel as a completely all-purpose instrument. An observer must therefore choose a telescope with his particular interests in mind.

By placing a telescope in an observatory the observer does more than provide protection for the instrument and himself. Perhaps more importantly, it makes it possible to have the instrumentation ready at all times at short notice. A perfectly clear sky suddenly clouding over can be very frustrating – frustration that is increased tenfold if it happens after half an hour spent setting up a telescope or camera in the open.

Astronomical observing should be an enjoyable pursuit, whether it involves making demanding observations or simply sky-gazing. A telescope properly set up in an observatory goes a long way to making astronomical observing even more of a pleasure.

I wanted to provide as much and varied information as possible, so this short chapter describes five markedly different observatories. Each contains very different types of instrumentation and, though extra detail of

Figure 4.1
H. B. Ridley's observatory in Somerset. The telescope is housed beneath a 12-foot diameter steel dome.

other systems is given where appropriate, they exemplify what is probably the most popular range of equipment employed by serious amateur observers.

H. B. Ridley's Observatory in Somerset*

This observer's mainly cometary interests require that he takes wide-field photographs using astrographic lenses mounted equatorially. A large refractor serves as the guide telescope. The whole system is housed inside a rotating dome with a wide observing slit. The large refractor and dome are recognisable as the 'classic' setup.

The 6-inch refracting telescope has a 90-inch focus. This delivers the high-quality image with good definition and contrast which one would expect when using a refractor. While this particular refractor has an old and slow *f*/15 doublet lens to minimise colour defects, *f*/8 and faster triplets and apochromats, which have superb colour balance, are now available. The short tube length of these new lenses is an attraction when a mount is to be considered. But it must be borne in mind that the image scale is correspondingly small. Slow refractors with large image scales are generally used for high-resolution work such as lunar and planetary studies and double-star measurement. It is much more difficult to achieve the high magnifications required for this type of work when using the new short-focus lenses. However,

---

* Sadly, since this chapter was written, H. B. Ridley has died and his observatory has been dismantled. – *Editor*

they come into their own when used for colour photography because of their improved colour correction and photographic speed. Both types can be used with CCDs for high-resolution imaging, and the typical apertures used by amateurs (4 to 8 inches) are well in line with the average seeing conditions experienced in the UK.

The main use of the refractor in this observatory is as a guide telescope for the wide-field lens when taking long-exposure photographs. Again, the long focus is advantageous because any slight deviation in the drive rate can be detected by the observer as he or she watches a guide star behind illuminated wires in the eyepiece. A refractor has generally been the favoured choice as guide telescope because of the stability of the optical components and the small, well-defined star images it gives. The guide telescope system has been completed

Figure 4.2
H. B. Ridley's 6-inch refractor and camera.

Figure 4.3
The up-and-over
shutter in the dome
of H. B. Ridley's
observatory.

by the addition of a filar micrometer to allow offset
guiding when a moving object such as a comet is being
photographed.

The dome of this observatory is of steel construction
and rotates on a circular angle-iron rail. The shutter is
of the 'up-and-over' type, and the internal diameter of
the dome is 12 feet. The internal diameter is an important dimension when considering the erection of an
observatory. To be able to work comfortably inside an
observatory plenty of space must be provided, and I
would recommend that 10 to 12 feet, either in diameter
or square, is the minimum size. The slit in the dome
needs to be at least 3 feet wide, though 4 feet would be a
better proportion for a dome of this size.

The siting of the equatorial mount within the dome
is important. Very careful consideration should be given
before pouring any concrete. Depending on the type of
mount and the tube length, placing the main support in
the centre of the dome may well be less effective when
perhaps offsetting it would provide more working
space. The height of the declination/altitude axis should
be above the bottom of the dome slit, and the dome slit
should extend past the zenith by a distance greater than
that of the telescope aperture. All these comments may
seem obvious, but it is possible to make elementary
mistakes. The installation of electric power should be
done by a qualified electrician and safety circuit-

breakers should be considered essential. Do not compromise on this point; it is not worth it.

The siting of a dome is a major factor when institutions are considering building a new observatory, but the average amateur is usually limited to selecting a good spot in the back garden. It may not be possible to have clear horizons in such situations, but try to arrange for the southern aspect to be maximised. Be very careful not to violate local planning regulations. Find out what is allowable before construction is begun. Local authorities can be ruthless when regulations are flouted, even to the point of having unauthorised buildings demolished. Dark skies, of course, are the most obvious consideration.

H. B. Ridley has chosen a particularly fine site for his wide-field photography, but not everyone can be as fortunate. Town sites – complete with artificial light – are not ideal, but even at these successful observing can be accomplished, as my next example proves.

Donald F. Trombino's Observatory in Deltona, Florida

Some amateurs suffer from a disease known as 'aperture fever'. The primary symptom is obvious – an unquenchable thirst for large telescopes. The prognosis is dim unless provision is made to properly house and care for these 'light buckets'. If exposed to the elements their fate is predictable – abandonment or, at best, eventual consignment to a dusty attic or damp cellar, never to see the light of day or the beauty of the night sky.

Building a first-class observatory is the obvious cure. The chances are you've invested a considerable sum in your 'scope and your remaining funds are limited. But cleverness and innovation are longstanding traits among amateur astronomers, and the resourceful will find many ways to keep construction costs to a minimum in order to protect their investment.

Not all of us can afford the luxury of living atop a tall mountain peak with pristine seeing for much of the year. For better or worse, your site – like the one described here – will likely be a corner of your garden or backyard. Your 'micro-climate' will play an important role in determining just how much you will see from your chosen location. It must be carefully situated away

from surrounding buildings and rooftops, including your own. A grassy location with low shrubbery will absorb rather than radiate heat. Avoid concrete or asphalt if at all possible.

The type of telescope you own and use will determine the best design for your specific needs. Our concept of the typical observatory is a circular, domed building with an opening protected by transverse, up-and-over or hinged shutters to protect the telescope when not in use. Domes are well-suited to Newtonian reflectors (especially open-tube designs), as well as to Schmidt–Cassegrain telescopes (SCTs). Generally they are costly, difficult to build, and usually have low entrance doors in order to give the telescope access to the horizon.

The elderly, handicapped, or tall individual who does not fancy getting down on all fours in the dark to crawl into an equally dark observatory would do better to consider an observatory of the roll-off-roof type. These are especially convenient for long-focus refractors. Their high walls permit full-height entrances, protect the telescope from gusts of wind, and give the observer an unobstructed view of the entire sky. Whether of single-piece or split, two-section, design, a roll-off-roof observatory quickly dissipates heat caused by the sun during the day, whereas domes can take several hours to cool off. In addition, heat-waves escaping through the dome's open slit can cause what is called 'dome seeing'.

Nothing in life is permanent. If you are comfortably retired or are otherwise absolutely certain that you will

Figure 4.4 The Davis Memorial Solar Observatory in Florida, owned by Donald F. Trombino and guarded by Ms Caspar, his cat.

stay put for a long time to come, a permanent observatory is the logical choice; if not, a modified, prefabricated garden shed is recommended.

Funds permitting, a well-built, permanent shelter is certainly preferable. Should unforeseen circumstances compel you to relocate at some future date, a well-thought-out, roll-off-roof observatory could be used for non-astronomical purposes (garden shed, workshop, sauna, studio, etc.). Believe it or not, there are people who do not share your enthusiasm for astronomy! Nevertheless, an æsthetically pleasing observatory can actually increase the value of your property, so appearance should be seriously considered in your observatory plans.

Central Florida's heat and humidity ruled out the use of brick or stone construction for Trombino's observatory, known as the Davis Memorial Solar Observatory. These materials tend to absorb heat during hot summer

Figure 4.5 Orion, seen above Donald Trombino's 6-inch solar refractor, through the opened roll-off roof of his observatory.

Figure 4.6 The interior of Donald Trombino's well-equipped observatory.

days and radiate it during the evening. Treated wood is less apt to cause unwanted turbulence above the observatory. The design chosen was a one-piece roll-off type with 8-foot-high walls to house a six-inch (150-mm) solar reflector. Power failures are a frequent occurrence during the rainy season, so the 1200-pound steel roof is manually operated using two 1400-pound capacity boat winches. A malnourished twelve-year-old could easily open and close the 16 ft × 18 ft roof! Besides, neighbours might not have appreciated a noisy, battery-operated winch during the small hours of the morning.

When fully opened, the heavy roof shades an attractive tropical patio on the north side of the observatory, giving visitors a comfortable environment while they await their turn at the telescope. Unsightly crossbeams, often used to reinforce track supports in some designs, result in wasted space. In this design not an inch of space is wasted, inside or out.

Despite the observatory's lowland location (only 80 feet above sea level) the waters of nearby lakes help stabilise the air above this observatory, making it especially suited to solar, lunar and planetary observing.

Florida is known as the 'Sunshine State'. In point of fact, it is the lightning capital of the world! Five thousand cloud-to-ground lightning strikes are typical during June, for example. Summer hurricanes with wind gusts of 120 miles per hour or more are not unusual. Transverse shutters of a dome (even in the closed position) could act as 'wind sails' under such conditions, with disastrous consequences! When the observatory is not in use, four heavy-gauge steel irons, attached to the walls and roof interior, securely bolt together to keep the roof locked in place. So far, they have weathered many severe storms without problems.

The 'unofficial' state flower of Florida is humorously referred to as the 'Fungus Floridicus'. It thrives on heat and humidity! To discourage this pest the interior walls of the Davis Memorial Solar Observatory are made of white plastic laminate similar to that used in low-cost bath and shower stalls. They are easy to maintain, never need painting, and can be wiped clean of mould and mildew (or fungus) with a liquid cleanser. Vents on all four walls and on the north and south portions of the movable roof provide excellent air circulation. Fungus does not like a well-ventilated room. White kitchen cabinets of the same material were installed above a 12-foot work bench. They conveniently store filters, cameras, CCD equipment and other large accessories. Fitted drawers contain a complete set of eyepieces, diagonals, and smaller filters. When they are closed, a small photocell automatically activates a 4-watt night light inside; this low heat source keeps the optics moisture-free.

The 12 ft × 14 ft room is windowless to cut costs and discourage vandals. A smoke detector and burglar alarm protect the equipment. As added insurance, a 'DANGER – HIGH VOLTAGE' sign posted on the out-

Figure 4.7 Donald Trombino adjusts the camera attached to his telescope.

side south wall of the observatory works wonders at keeping unwanted night guests away!

The open roof sometimes invites surprise visitors – an occasional squirrel, curious racoons, a mocking bird, a friendly Florida salamander or other herptiles (Greek 'herpeton', meaning 'creeping thing') for which Florida is so famous. These creatures are harmless and certainly add an element of excitement when they decide to 'drop in'.

It has been said that 'A friend in need is a friend, indeed'; especially if those friends happen to be amateur astronomers by night and skilled welders, carpenters, electricians, cabinet installers, bricklayers, etc. by day! In exchange for observing time, their outside talents will considerably reduce your overall construction costs.

However, if you *must* rely on commercial building contractors, take time to check their credentials. A few common-sense precautions will avoid grief further down the line.

- *Experience.* How long have they been in business? Have they done work in your area, or are they merely 'visiting', never to be seen again?
- *Affiliations.* Is your contractor affiliated to a professional society, and insured and licensed? Ask for written proof!
- *Availability.* Like a hit musical, good builders are usually booked months in advance, especially during the busy spring or summer months. Beware of the builder who is ready to start at the drop of a hat!
- *Referrals.* Ask for the names and addresses of previous customers. Has the contractor done work for your neighbours? If so, was the work satisfactory?
- *Demand a contract.* The contract should clearly stipulate labour and materials costs, and what work is to be performed or not to be performed. Insist upon a firm completion date in writing.
- *Payments.* Beware of demands for cash payments in advance for any work or building supplies. Reputable contractors have well-established lines of credit with banks and sub-contractors, so there should be no need for 'up front' money from you!
- *Reputation.* Always check beforehand with your local consumer protection agency, Better Business Bureau, or national equivalent. Although they will not recommend contractors by name, most will at least report the number of complaints against your candidate. The general rule here is *caveat emptor*!

In addition to daily solar observations in white, Hα and calcium-K light, Trombino's instruments are used to evaluate new equipment designed for solar observers. The Davis Memorial Observatory is often made available to students and serious amateurs for research projects. Presently it is the *only* fulltime optical solar observatory in Florida. The rewards derived from freely sharing it with others are indeed gratifying.

Anyone wishing to obtain detailed plans for a roll-off-roof, dome, or 'clam-shell' type of observatory should write to: John Hicks, PO Box 75, Keswick, Ontario LP4 3E1, Canada.

M. Mobberley's Observatory in Chelmsford, Essex

This observatory is situated at a town site affected by sodium street-light pollution. The telescope is a large, fast reflector (19-inch $f/4.5$) with a new CCD camera. The observatory is of the run-off shed type and the setup has been installed in a small back garden surrounded by houses. This is a modern instrument in every sense.

In general, large reflectors (12 to 24-plus inches) have become standard instruments for amateurs and have been employed for the full range of observing. The various configurations (Newtonian, Cassegrain, etc.) can all be found in amateur hands, and when well made perform excellently. Unfortunately there are also some poor examples available, and a check on performance needs to be undertaken before a full purchase is made. Don't get caught out! For visual observing. the large-aperture reflectors allow faint objects to be seen and high resolution to be achieved. In recent years these telescopes have been used by photographers for long exposures, and some remarkable results have been accomplished. The introduction of new photographic emulsions has allowed amateurs to record very faint objects. When CCDs are used on these telescopes the results can be breathtaking.

This particular telescope was designed for one main purpose: the imaging of very faint objects. The main mirror is a fast $f/4.5$ of 19-inch aperture, and has been made from the latest low-expansion glass. Its German equatorial mounting is fitted with a very precise drive which employs a large, accurate worm wheel with a stable electronic controller. At the focus is a CCD camera with a chip size of 6 mm × 4.5 mm. The resulting small field of view (about 8 arcmin) is very restricted compared with the half-degree field available when using the same instrument with 35-mm film. Where the advantage of CCD over film is evident is in its capacity to reach magnitudes fainter than 20 in reasonably short exposure times which would not have been possible a few years ago. This allows the monitoring of the faint minima of variable stars, especially cataclysmic types such as novæ and supernovæ. That this can be done at a light-polluted site near the centre of a large town is remarkable. The progress in this area of imaging gives the lie to the belief that successful observing must be conducted from a dark rural site. This should encourage all amateur astronomers.

The simplest form of observatory is a covering which can be removed completely when the observing session begins. Such is used here and works well. This design is probably the most cost-effective available. The structure is made entirely of wood and resembles an ordi-

Figure 4.8
M. Mobberley's observatory, run off to reveal his 19-inch telescope on its German equatorial mount. Note the rails on which the wooden structure travels.

Figure 4.9
M. Mobberley's
19-inch reflector
'parked' within his
observatory.

nary garden shed. The whole building is just big enough to fit over the telescope and runs off on rails at ground level. Electrical installation is via underground conduits providing power at the telescope pier. While the advantage of weather protection during observing is absent, this small back-garden site is less affected by adverse conditions than an exposed rural one would be.

Dame Kathleen Ollerenshaw's Observatory in Cumbria

This newly constructed observatory is situated in a very dark site in the Lake District and has been built to blend with the surroundings. It contains a 10-inch Schmidt–Cassegrain telescope which is fitted with a CCD camera. The owner's main interest is in recording deep-sky objects.

During the 1970s and 1980s the amateur's usual choice of telescope, the reflector, was joined by a newly developed family of catadioptric systems which were marketed world-wide from the USA. The most popular was the Schmidt–Cassegrain and, to a lesser extent, the Maksutov–Cassegrain. These telescopes were promoted as being portable and powerful. The folded design of the systems meant that they provided long focal lengths ($f/10$ and greater) and high magnifications. In order to appeal to photographers and observers who wished to use these telescopes for deep-sky work, the early fork mountings were improved and a focal-length-reducing accessory was introduced. Drive systems were also

upgraded and are now very sophisticated. These telescopes have become effective systems and are now able to be adapted for use with computer control.

One of the most difficult operations for amateurs seems to be pointing the telescope at a designated area of sky. This can prove even more problematical if that sky area is in a barren region or if the object of interest is too faint to be seen in the finder. These technologically advanced telescopes overcome this problem for observers by employing shaft encoders on both rotation axes of the telescope. A simple star-alignment procedure informs the drive's computer where the telescope is pointed and new positions are found automatically by a 'GOTO' facility. This is linked to a database which may contain many thousands of object positions. If the equatorial fork mounting is correctly polar-aligned then this facility works with impressive speed and accuracy, allowing faint objects to be readily found. This allows extra time for observing the object rather than searching for it. This is an obvious advantage when a series of fields or objects need to be observed, as in variable star work or supernova searching. Other features of such telescopes include motorised focusing, smart drives with autoguiding, and the capacity for linkage to a personal computer. This in turn allows the possibility of remote observing. The introduction of computer technology into the control of telescopes is leading to many welcome and radical changes to the observing practices of amateurs. Recently introduced are large-aperture cata-

Figure 4.10 The roof of Kathleen Ollerenshaw's observatory rolled back ready for use.

dioptrics (16 inches and greater) on altazimuth mounts with very sophisticated electronic controls. These set a new standard in telescopes available for amateurs.

Kathleen Ollerenshaw's observatory contains a working example of one such telescope. It is a 10-inch Schmidt–Cassegrain with an automatic slewing facility and a large database; consequently the telescope can be slewed to any position which the observer keys in. The drive is a 'smart' system which corrects for any periodic errors present in the primary gear train. This reduces tracking errors to a few arc seconds and is good enough to allow the observer to take wide-field photographs with a camera 'piggybacked' on the main telescope (see Chapter 10) without manual correctional adjustments being made. When photographs or CCD frames are taken through the main instrument the drive is controlled by a separate CCD autoguider. This is connected either to the main telescope via an off-axis guider or to a small auxiliary guide telescope. The autoguider makes small corrections to the drive every second or so after registering a guide star drifting between pixels on the matrix of the CCD. To see the telescope move quickly between NGC galaxies, and see the images appear one after another on the CCD monitor, is impressive. Given several clear nights, a significant number of objects could be recorded with relative ease. After many years of hard-working attempts to obtain good long-exposure photographs, new-technology telescopes such as this are a revelation.

The design of the observatory is of the roll-off or run-

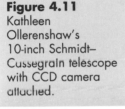

Figure 4.11
Kathleen Ollerenshaw's 10-inch Schmidt–Cassegrain telescope with CCD camera attached.

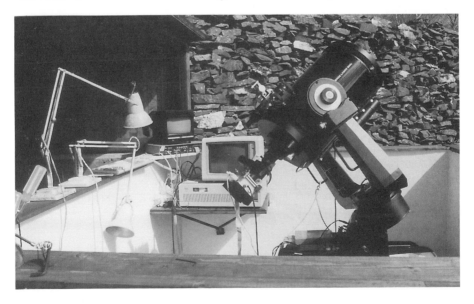

off-roof type. The roof section is made of tongued and grooved boards and runs off towards the north on angle-iron rails and rubber wheels. It is very heavy, and employs a cleverly designed pulley system which is operated by a bicycle gear and chain. The whole roof moves as one section. The south end parts centrally into two separate doors, each of which hinges back. The roof and doors can be partly opened in combination to provide some protection from the wind during observations. Electricity and telephone lines enter the observatory via underground conduits. The walls have been constructed of the local slate stone, as has the telescope pier. This type of observatory is effective in use and easy to build.

My own Observatory near Lancaster

This observatory contains a 21-inch (530-mm) *f*/5 Newtonian reflector in a 23-foot (7-m) diameter dome. The telescope is principally used for recording comets and eruptive variables. It is situated in a dark rural site on land at the rear of my home.

The 21-inch telescope is supported on a large cross-axis mounting. This mounting also carries a 6-inch (150-mm) *f*/19 refractor, an auxiliary 12-inch (300-mm) *f*/7 reflector and a 5-inch (130-mm) wide-field Schmidt–Newtonian. All these instruments can be used visually

Figure 4.12 The 23-foot diameter dome of Denis Buczynski's observatory.

Figure 4.13 Denis Buczynski's 21-inch reflector on its English cross-axis mounting.

and for photography and CCD imaging. An intensified TV camera can also be attached for continuous recording. The whole instrument is driven by a 30-inch (0.75-m) diameter friction drive at the south end of the polar axis. Cabling for the cameras on the telescope feed to an enclosed workstation situated northwards of the north pier. Guiding can be done visually through the 6-inch refractor. It can also be done viewing a monitor which is fed by the intensified TV camera attached to the 12-inch. The acquisition of fields must still be achieved manually using setting circles, but the observing can be done from a sitting position at the workstation desk. The instrumentation has evolved over 10 years, rather than having been planned as such. As telescope technology has advanced and become more affordable, items of equipment have been added to improve observing capacity. The advantage of having many telescopes and lenses mounted together allows a choice of whichever size of field is to be recorded. With the CCD the range is from 7 arc minutes to over 4 degrees.

Figure 4.14
Control terminals for
Denis Buczynski's
CCD and TV
cameras.

The dome is a converted farm silo made from gal-
vanised steel rotating on a 2-inch circular pipe rail at the
level of the suspended wooden floor. The shutters are
bi-parting and the observing slit is 6 feet wide. The
overall height of the observatory is 20 feet and this size
did need Local Authority planning permission.
Electrical power is supplied via buried armoured
cabling. The 23-foot diameter silo top is the largest,
although 16- and 12-foot versions are also manufactured. The cost of these, when new, is prohibitive.
However, if you can find a farmer who is about to
demolish a silo tower then it can become a very cheap
option. As such domes were made to be positioned at
the top of 100-foot towers they are manufactured to
withstand extreme weather conditions. The interlock-
ing structure makes fabrication simple and watertight.
The slit needs to be cut out and shutters fitted, but this
need not be too great a problem. The rotation of the
dome is achieved by using V-wheels attached to the
dome base. These run on a circular steel pipe fixed to

the observatory wall. After ten years this observatory has proved excellent in use and has caused no real problems, and on that basis I can recommend this design.

In conclusion it must be said there are many variations on these styles of observatory. The range of telescopes available also includes Dobsonian reflectors, Schmidt cameras, and old telescopes such as the large Victorian reflectors and weight-driven refractors. These last should not be regarded as outdated curiosities but as classic instruments, to which the addition of a CCD camera would perhaps be the ultimate blend of old and new technology. Telescopes uniquely extend our limited vision of the Universe. Enjoy every view through them.

Chapter 5

Making Your Own Telescope

John Watson

Modern astronomical telescopes represent very good value for your money. There are a few notable exceptions (see Patrick Moore's remarks in Chapter 2), but in general they are good value. This doesn't mean that they are cheap, however.

To many enthusiasts an astronomical telescope of reasonable size represents a major outlay that may be hard to justify in the domestic budget when stacked up against the cost of keeping a car on the road, the money involved in keeping the house watertight and decorated, the price of the occasional holiday, and the sometimes unbelievable expense of rearing one or more children.

The 'do-it-yourself' option is one that few people consider – but I successfully manage to assemble a bedroom full of flat-pack furniture, install a shower in my bathroom, and build my son a model glider; so why not make a telescope? If you're happy to make a fairly simple instrument – to forego electric drives, slow-motion controls and the like – it's not in the least difficult. There's no reason why you can't put in all the extras, but it depends on your abilities as a carpenter, engineer, or builder of improvised machinery.

Useful piece of advice number 1. If you are a newcomer to telescope making, keep your home-built telescope simple, or you'll probably never finish it. Save the grandiose plans for later.

There is certainly insufficient space here to go into much detail about how to build *any* kind of telescope, but what I will try to do is to give some guidelines for newcomers, and a few tips for those who have already had a go at making their first telescope.

Reflectors Versus Refractors

Most home-made telescopes are reflectors. Aperture for aperture, reflectors are substantially cheaper than refractors. Why? Well, a telescope mirror has to have one surface that is accurately ground and polished. The glass (or ceramic) that it's made from doesn't even have to be transparent, because the top surface is aluminised and light doesn't pass through the mirror at all. In contrast, the two-element objective lens of a refractor has *four* surfaces that have to be ground and polished, and the lenses have to be highly transparent, homogeneous, and of the proper refractive index. The two elements have to be mounted in exactly the right positions relative to each other, which necessitates a precision-made, turned metal 'cell' to hold them.

The reason for 'doing it yourself' is usually cost, so it follows that the cheaper option of a reflector is the one that most of us opt for. The tube and mounting are easier to make, too. Everything I am going to say about reflectors here applies to *Newtonians*. Cassegrains (and other designs) are nowhere near as easy to make.

Making Your Own Optics

Useful piece of advice number 2. Think carefully before making your own mirror.

I admit it, I'm prejudiced. In my opinion, to make a Newtonian primary mirror (the easiest place to start) you need to be very patient, retired and with plenty of time, or happy with working to very long time-scales – preferably all three. I enjoyed making my telescope mirror, but I also ended up wondering why I hadn't done some odd jobs, freelance work, or whatever it would have taken to earn enough money to buy something professionally made.

There is insufficient space here in this book full of different aspects of astronomy to go into detail – to give you a recipe for making a telescope mirror. What I hope to do is to give you the flavour of what is involved, rather than a detailed description. If it is to your taste, then you should get hold of one or more books devoted to the subject of making astronomical optics. The two that are most highly recommended are Neale E. Howard's *Standard Handbook for Telescope Making* and *Making Your Own Telescope* by Jean Texereau (see the Bibliography at the end of this chapter).

It is usually recommended that an amateur should start with a 6-inch (150-mm) or 8-inch (200-mm) mirror of around *f*/8, on the basis that anything bigger or with a larger relative aperture is noticeably more difficult – and expensive should it all go wrong – and anything smaller is not worth doing.

That advice is no doubt sound, although the first (and only) mirror I made was a 4-inch (100-mm) *f*/8, probably on the grounds of pessimism. It actually isn't that difficult, but in these hectic times when hours count and I get irritated if I have to wait more than a couple of seconds for my word processor to do a spell check, grinding a mirror seems to take a long, long time.

What You Need

To begin, you need two Pyrex or similar glass 'blanks', discs of glass that are the right size and thickness. You don't really have to have low-expansion glass for the blank that is to be the 'tool' rather than the mirror, and this can save money.

Blanks for a small mirror (up to 8-inch) are reasonably inexpensive, but you'll probably be astonished at how much you have to pay for the blanks to make a 10-inch (250-mm) or bigger mirror. You'll also need a quantity of fine pitch, several grades of abrasive for the grinding, and a polishing compound. These materials aren't to be found in your local hardware store, but fortunately there are a number of suppliers who can let you have (by mail if necessary) a complete 'kit' for making a mirror.

How to Start

First, you need to know what it is you are trying to make, in terms of focal length. It's quite hard to arrive at an exact focal length unless you're experienced – at least, I found it difficult – so make the mirror before you make the telescope tube!

Construct a template of the curve you're aiming to produce, using thin plywood (1 or 2 mm thick, from the model shop). Draw the curve using a long batten with a pencil jammed into a hole drilled through one end and a nail banged through the other end into the bench. It sounds crude, but it works. The distance between the nail and the pencil is the required radius of curvature, i.e. twice the focal length. Cut the ply with scissors, and use a file to get it as accurate as possible.

The concave curve of the mirror's surface is made – 'generated' is the word used – by grinding the mirror and tool against each other, with a mixture of abrasive and water in between them. The tool has to be solidly fixed to a (very) solid surface. I used a piece of 5-mm ply with a 4-inch hole in the middle into which the tool could just be fitted. I then nailed the ply to a workbench. The tool can be prised out if need be.

Grinding

You put some of the coarsest abrasive – mixed with water – between the glass discs and get grinding. Moving the mirror (the disc you're holding) back and forth across the tool in more or less random strokes, rotating it a little, will cause the mirror gradually to become concave and the tool to become convex. (Why? It's because, if you apply even downward force to the mirror, the weight is spread over a smaller area when it overlaps the tool. This produces more pressure on the edge than on the middle of the tool and vice versa on the mirror.)

Keep going, washing everything off and replenishing the abrasive and water whenever you can feel the cutting action diminishing. Use the template to keep checking the curve, and when it's about there, begin to work your way through progressively finer and finer abrasives (usually four or five different grades).

Different strokes of the tool give slightly different effects. Your mirror kit or book will go into detail.

As you reach the shape of the final curve you must use one or more of a variety of optical tests to determine the focal length and accuracy of the curve you are generating. The accuracy of the curve is much more important than the precise focal length, of course. Modifying your stroke pattern will correct most problems, but it is essential to keep your work perfectly clean, especially during the fine grinding stages. A piece of grit (or left-over coarse abrasive) will make a scratch that is very hard to eradicate.

Although it didn't happen to me, the mirror and tool sometimes suddenly stick together – although if you wash off and replenish the abrasive often enough, it doesn't seem to happen much. Getting the two pieces of glass apart again can be difficult, and procedures involving a slow escalation in violence are usually recommended, starting with putting the discs in the refrigerator, through running them under a hot tap, and ending with the last resort of laying into them with a wooden mallet.

Polishing

Polishing is done with a pitch 'lap', which replaces the glass tool. There are a number of ways to make the lap, but I simply fixed a cardboard rim round the edge of the mirror (like a cake tin) and poured hot pitch into it to a depth on an inch or so. Do this carefully: it's messy, and hot pitch is very sticky and thus quite dangerous, because it can cause severe burns if you get it on your skin.

You have to cut slots in the lap, and the shape of the slots, along with kind of strokes you make, determines the final figure of the mirror: it has to be parabolic, remember. The actual polishing is similar to grinding, except that it's much harder work, and you have to keep checking the figure of the emerging mirror, using optical tests. You usually polish the mirror to a spherical surface before 'parabolising' it. The difference between a spherical curve and a parabolic one is critical but not great.

Finishing

Once the mirror is finished to your satisfaction, you need to have it aluminised. You can't do this yourself. The company that supplied your mirror kit (or abrasives, etc.) will normally be able to get your mirror aluminised, otherwise look through the pages of the astronomy magazines.

My first little mirror actually performed reasonably well. I still have it – and last used it in a hastily constructed guide telescope to help me photograph the last – disappointing – appearance of Halley's Comet.

Buying the Optical System

It is possible to buy new or secondhand primary and secondary mirrors. A good primary mirror will be made of a low-expansion glass or ceramic; avoid plate-glass mirrors because they take hours (literally) to settle down when moved from a warm environment (indoors) to a cold one (outdoors). Pyrex is the commonest kind of low-expansion glass, but various ceramic materials – for example Cervit or fused quartz – are used in some of the best mirrors.

You will find mirrors described by their surface finish; $^1/_4$-wave – flat to within a quarter of the wavelength of light (green, usually) – is sufficient for all normal purposes; it isn't necessary to look for higher specifications. However, look out for manufacturers' hyperbole: as with the power output of a hi-fi amplifier, it all depends how you measure it, and ' $\pm^1/_{16}$-wave parabolic' actually translates into ' $^1/_4$-wave on the wavefront', which is what Rayleigh meant when he was defining the theoretical requirements for a primary mirror.

The aluminium coating on a mirror will last between two and ten years on average, less in a city and more in the country where the air is cleaner. A reputable seller of secondhand mirrors will replace the coating of a 'previously owned' mirror as a matter of course before he sells it.

Useful piece of advice number 3. Remember that you can't tell how good or bad a telescope mirror is by looking at it, or by the price asked for it, especially secondhand. Buy from a reputable source – preferably one with the facility to test the quality.

It is 'common wisdom' that the thickness of a primary mirror ought to be one-sixth of its diameter. This is probably over-cautious, and you shouldn't be put off by a mirror which is thinner than this, say one-eighth of the diameter.

Newtonian secondary mirrors have to be *flat*. The surface needs to be at least as good as that of the primary. Don't skimp, buy a new one (as opposed to secondhand) if you have any doubts. It's worth the extra money. The same applies to eyepieces.

*Useful piece of advice number 4. Buy a bigger mirror
than you think you need or can afford, otherwise you'll
end up making two telescopes.*

The Telescope Tube

The tube of a reflector can be either closed or a 'skeleton'. There are pros and cons for each and, as is often
the case when there are firm protagonists for opposing
notions, the differences are in practice slight. I've made
and used both, and the results (even photographically)
haven't been noticeably different. There is also a lively
debate about whether it is best to make the tube out of a
thermally insulating material like wood (so that your
body heat won't warm part of the optical path and cause
distortions) or a conducting material like metal (to
achieve thermal equilibrium more quickly). Both seem
to work.

The telescope tube has three purposes. First, it holds
the mirrors and eyepiece rigidly in position in relation
to each other. Second, it shields the optical components
from skylight and, to some extent, dew. Third, it stops
you crashing into the optical components in the dark!
This last point is not as silly as it may sound: paint the
inside of the tube matt black to avoid reflections, but
paint the *outside* white to avoid collisions.

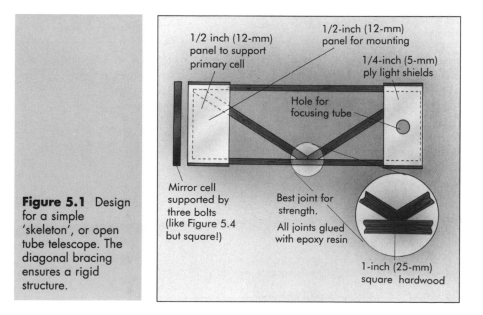

Figure 5.1 Design
for a simple
'skeleton', or open
tube telescope. The
diagonal bracing
ensures a rigid
structure.

1/2 inch (12-mm)
panel to support
primary cell

1/2-inch (12-mm)
panel for mounting

1/4-inch (5-mm)
ply light shields

Hole for
focusing tube

Mirror cell
supported by
three bolts
(like Figure 5.4
but square!)

Best joint for
strength.

All joints glued
with epoxy resin

1-inch (25-mm)
square hardwood

It is probably easiest to make the telescope tube out of wood, because for most of us it's easier to work than metal or plastics. Cardboard is pretty good, too. You can get cardboard tubes from all sorts of places – the core of a roll of carpet, for example (up to six inches) or from some of the larger builders who use them for casting concrete pillars (as big as you're likely to need). Figure 5.1 shows a simple square 'skeleton' tube.

For my most recent telescope (a 12-inch) I used a form of construction borrowed from the model glider I made for my son many years ago. The design is illustrated in Figure 5.2. I first cut six 460-mm (18-inch) diameter discs out of 15-mm plywood, then cut 355-mm (14-inch) diameter holes in the middle, leaving 52-mm (2-inch) thick rings. I drilled ten 12-mm (½-inch) equidistant holes to take beech dowels, which made the basic structure. Then I filled in all the gaps (forty-nine of them!) with sheets of 4-mm balsa wood (from my local model shop) to make the structure rigid and provide light-shielding. Balsa is an excellent insulator, weighs very little, and is extremely easy to work with. The resulting tube, shown in Figure 5.3 after painting, is strong, lightproof, very rigid, and weighs just 5.5 kilograms (around 12 pounds).

It is easy to attach the mirror cell (see below), and the many flat panels make it easy to mount accessories (a guide telescope, for example). Also, it looks nice!

Useful piece of advice number 5. Check the focal length of your primary mirror very carefully, and draw a plan of

Figure 5.2 Design for a built-up closed tube telescope, using construction techniques borrowed from aircraft modelling.

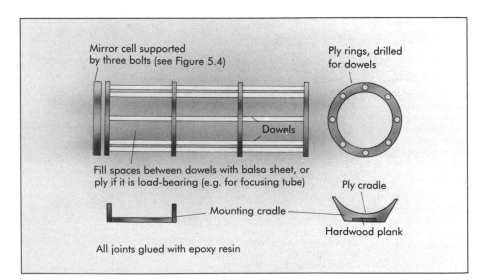

Mirror cell supported by three bolts (see Figure 5.4)

Ply rings, drilled for dowels

Dowels

Fill spaces between dowels with balsa sheet, or ply if it is load-bearing (e.g. for focusing tube)

Ply cradle

Mounting cradle

Hardwood plank

All joints glued with epoxy resin

Figure 5.3 The author's most recent telescope, the design of which is shown in Figure 5.2.

the whole telescope, either full-sized or to an exact scale. It's really depressing to spend weeks making a telescope tube that is the wrong length.

The Mirror Cell

A lot has been said about mirror cells. The idea is to hold the mirror firmly in place without bending it, regardless of the angle of the telescope. You can buy a 'multi-point suspension' cell that supports the back of the mirror at several carefully calculated points, but such cells are expensive and may be difficult to attach to a wooden tube. Here's an alternative that can be made with no more than a power jigsaw (sabre saw) and a drill.

The design I used – and which has worked well for a couple of years – has the mirror resting on a mat of 'bubble wrap' polythene sheet. Designed for protecting fragile items in the mail, bubble wrap has lots of partially inflated air bubbles in it and provides a very effective multi-point suspension (hundreds of points, in fact!) when placed between the mirror and a stout wooden backplate. Any uneven pressures are balanced by the bubbles.

I made the backplate out of two thicknesses of 12-mm marine-quality plywood, stuck together with resin woodworking adhesive. The upper thickness is circular, while the lower one is in the shape of a three-armed cross. Each arm is drilled (through both discs) for the three bolts that will support the cell and attach it to the telescope tube.

Two semicircular clamping rings were cut from 6-mm ply, with notches to leave room for the cell support bolts. The usual builder's three-ply is coarse and splintery; you need five-ply, which can be cut precisely. Timber yards may want to sell you a huge sheet of it, but model shops will have smaller pieces. The clamping rings should be carefully filed so that they are a fairly snug fit round the mirror, within 0.5 mm all round.

Pieces of 25-mm batten – ten in all – support the clamping rings about 5 mm from the upper surface of the mirror. Brass woodscrews hold the clamping rings, through the pieces of batten, to the cell base. Once the cell is finished and painted (matt black for the upper surfaces of the clamping rings), it is ready for assembly.

Cut out a suitable disc of bubble wrap, place it on the cell, and leave the mirror on it for a few days to let the bubbles flatten out and settle down under the weight of the mirror.

Then screw one clamping ring in place and cover the inside edge (where it touches the mirror) with a thin layer of silicone aquarium adhesive. This is a very strong adhesive used for making fish-tanks; don't use any other sort of silicone adhesive, because it won't be strong enough. Put the mat of bubble wrap in place, and carefully push the mirror up against the clamping ring. Put some adhesive on the second clamping ring, push it

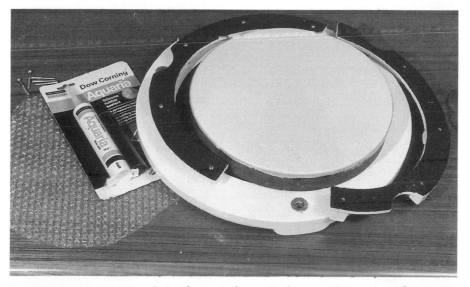

Figure 5.4 A wooden mirror cell which uses bubble-wrap to support the 12-inch mirror.

into place, and put in the securing screws. Leave the whole thing to dry before moving it.

The cell works well and is easy to make. It is very secure: although the rings are only 6 mm thick there is a total area of about 5000 mm² of adhesive. If you want to take the mirror out (for re-aluminising), just undo the clamping ring bolts and peel the clamping ring away from the glass.

The complete but unassembled cell, along with a piece of bubble wrap and the adhesive, is shown in Figure 5.4.

It is important to have a close-fitting mirror cover: I used a plastic saucer that was meant to stand under some enormous plant pot.

Secondary Mirror Supports

The Newtonian secondary mirror has to be held in the light path very firmly by some structure that blocks the minimum amount of light. The support, called a 'spider', typically consists of four thin blades supporting a turned metal holder that contains the secondary mirror. The idea is simple, but in practice it is complicated by the fact that the mirror has to have a mechanism that allows you to make fine adjustments to its angle, so that you can align (collimate) the telescope's optical system.

Figure 5.5a and **b**
The secondary mirror
(top) is supported by
a 4 × 40-mm steel
blade. Three screws
at the top of the
mounting tilt the
mirror for collimation.
The mirror support
assembly and the
focusing tube are
all mounted on a
single removable
rectangular plate
(bottom).

You can buy ready-made spiders and mounts, and
depending on your metalworking skills and tools you
may want to choose this option.

My own telescope has a slightly different approach,
the secondary being supported on a single, relatively
stout blade. For amateur telescopes this is quite viable,

and means that the whole 'viewing assembly' – the secondary mirror with its support, and the focusing tube and eyepiece – can be fixed to a single panel of the telescope tube – see Figure 5.5 a and b.

Focusing Tubes

There are two sizes: 1¼-inch and 2-inch. The larger size is better because it will accommodate wide-field eyepieces and is better for prime-focus photography. It probably isn't worth making your own focusing assembly unless you're an expert metalworker.

Collimation

Aligning the optical system of a Newtonian telescope is essential and must be done very carefully. It is quite easy, however, and unless you're going to use very high magnifications, rather less critical than many books would have you believe. The telescope design I've described here makes collimation fairly convenient, although you may need someone to help if your telescope is long enough to mean that you can't reach the mirror cell while looking through the eyepiece.

Stage one is to get the primary in the right place relative to the focusing tube. The prime focus should coincide with the focus of the eyepiece, with the focusing adjustment set about halfway along its travel. With the eyepiece removed, look down the focusing tube and adjust the three bolts holding the cell until your view of the top of the tube looks more or less concentric with the primary mirror.

Now you need to see where the focal point is. The easiest way is to remove the focusing tube, point the telescope at the Moon (*never* the Sun!) and use a piece of ground glass (or tracing paper) to see where the image is by projecting the picture onto it. Adjust all three cell support bolts together until you've got the focal point where you want it.

You now have to get the secondary, primary, and focusing tube all properly aligned. The way you make the adjustments will depend on the secondary support and the spider design, but the objective is to get everything looking concentric when you look down the (fully

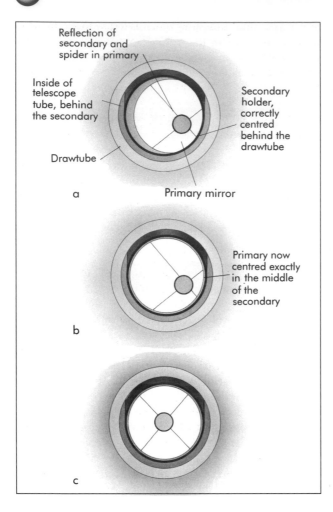

Reflection of
secondary and
spider in primary

Inside of
telescope
tube, behind
the secondary

Secondary
holder,
correctly
centred
behind the
drawtube

Drawtube

Primary mirror

a

Primary now
centred exactly
in the middle
of the
secondary

b

c

Figure 5.6
Collimating a
Newtonian:
a neither primary
nor secondary
aligned; **b** after
aligning the
secondary; **c** after
aligning the primary
so that the reflection
of the secondary is
exactly concentric
within it.

extended) focusing tube with no eyepiece in place. Adjust the secondary, then the primary. Keep on repeating the adjustments until everything looks right. You should end up with a view more or less like the one shown in Figure 5.6c.

Useful piece of advice number 6. If you think the collimation is nearly right, try it out! Then re-adjust your telescope optics more carefully and try it out again.

Mountings

Whole books have been written about making telescope mountings, and I won't attempt to go into much detail here. If you want something quick and easy to set up, why not start with a Dobsonian? It is possible to make one quite quickly, using only plywood sheet and the trusty jigsaw. Dobsonians are idea for telescopes with square tubes. Figure 5.7 shows the general idea. Obtaining the pieces of Teflon may be the only problem, but suppliers of telescope parts (the place where you bought your focusing tube!) will usually sell it in strip form.

It's important with Dobsonians that the telescope balances exactly at the horizontal axis. With a fairly lightly constructed tube this balance point will be surprisingly near the bottom of the telescope, so the Dobsonian mounting will be quite compact.

Equatorial mountings are much more difficult to design and build, and you should do plenty of reading before you begin.

Bob Turner FRAS, of West Sussex in the south of England, describes a simple equatorial mounting that can be made out of iron pipe fittings.

Figure 5.7 Design for a Dobsonian mounting.

One of the best ways that I have found over the years to produce a very cheap but very stable equatorial mount is to make the system out of black iron gas plumbing fittings [see Figure 5.8].

By using gas fittings in this way, and having bearings that are running on the pipe threads themselves, telescopes up to 10 inches aperture can be supported

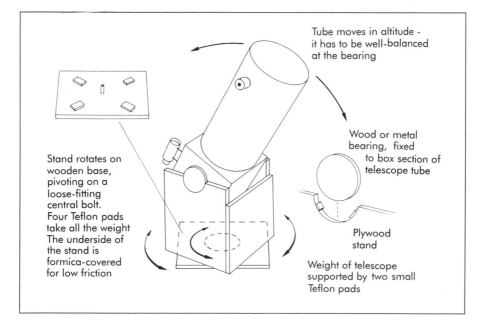

Tube moves in altitude - it has to be well-balanced at the bearing

Wood or metal bearing, fixed to box section of telescope tube

Stand rotates on wooden base, pivoting on a loose-fitting central bolt. Four Teflon pads take all the weight The underside of the stand is formica-covered for low friction

Plywood stand

Weight of telescope supported by two small Teflon pads

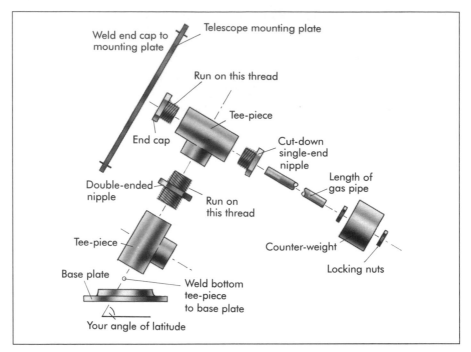

Weld end cap to mounting plate

Telescope mounting plate

Run on this thread

Tee-piece

End cap

Cut-down single-end nipple

Length of gas pipe

Double-ended nipple

Run on this thread

Tee-piece

Counter-weight

Base plate

Weld bottom tee-piece to base plate

Locking nuts

Your angle of latitude

on a stable equatorial mounting for less than £50, and smaller mountings can be even cheaper. I recently made a mount for a 3½-inch refractor out of one-inch pipe fittings for £12. About the most serious problem with this kind of mounting is the difficulty of fitting the driving mechanisms if you want to upgrade it.

Once again, as with telescopes, in mountings you only get basically what you pay for, and the more sophisticated and better manufactured the item, the more it is going to cost.

Throughout all my experience of over 40 years of telescope making, I have always found the old adage that 'you cannot make a silk purse out of a sow's ear' to be something that is very true, although countless amateur astronomers will try to 'make do' to their subsequent disappointment.

Traditionally however, the DIY telescope market has been something that has provided enormous amounts of pleasure to many people, allowing them to make something that performs well. In order for this to happen, however, engineering skills are required – as well as a better than average tool-kit, and the occasional use of engineering facilities to manufacture the parts that cannot be made at home.

Figure 5.8 Design for an equatorial mounting made from gas fittings.

I think Bob Turner's excellent drawing can't be bettered as a plan for this kind of mounting.

Don't be Disappointed

My first home-built telescope was a 2¹/₄-inch refractor made from an army-surplus lens of about 24 inches focal length. I was delighted by its performance. My first reflector was a 6-inch, and I was equally astonished by how well it performed.

The last telescope I made was the 12-inch described here, and at 'first light' on a crystal-clear cold night its performance was very disappointing. I checked and rechecked the collimation, and eventually came to conclusion that I had bought – from a normally reputable source – a rather poor primary mirror.

A few nights later, I tried again. The weather wasn't as good: the sky looked slightly misty. But the results were incredibly better! I had forgotten that large apertures are much more affected by 'seeing' conditions, and that a night on which the stars twinkle down out of a jet black sky is not necessarily the best kind of night for a still atmosphere.

Bibliography

Howard, Neale E, *Standard Handbook for Telescope Making*, Harper Collins, New York (1984)
Texereau, Jean, *How to Make a Telescope*, John Wiley, New York (1957)

Chapter 6
Auxiliary Equipment

M. Mobberley

Amateur astronomy is a relatively rare pastime when compared with more common pursuits such as golf or rock climbing. The amount of money involved in this hobby is therefore correspondingly small. Nevertheless, in the last twenty years a number of companies specialising in astronomical accessories have emerged (mainly in America) to cater for the more advanced amateur astronomers. The vast majority of these accessories are associated with the art of astrophotography, but visual observers are also well catered for. Some of these accessories would be hard for even the skilled amateur to construct by himself; others can be built by the resourceful, or more importantly, patient, amateur.

For the serious observer, it is essential to determine what astronomical field he or she wants to specialise in and then determine what auxiliary equipment is required.

As with any purchase, it is always a good idea to seek the opinion of others who may have more experience than yourself. The best people to consult will invariably be those who are known experts on a national basis, i.e. the most skilled observers and/or photographers.

Some of the more expensive astronomical accessories are reviewed in the US magazine *Sky & Telescope*, and these reviews are well worth reading for a frank assessment of a potential purchase.

Because of the bewildering array of accessories available, the only way to tackle this subject is to split it into categories and deal with the more popular items in each category in some detail.

Astrophotography

Photographic Equipment

This subject is dealt with more fully in Chapter 10. However, even at the risk of duplication, a summary of relevant equipment will do no harm.

At its simplest, astrophotography consists of an undriven 35-mm camera on a fixed tripod. Such an arrangement can easily capture bright meteors during the major shower periods. At its most complex, astrophotography can involve photographing sixteenth-magnitude comets, allowing for sidereal and apparent cometary motion and carrying out astrometric measurements on the negatives for determining the comet's orbit. Various levels of equipment are available to assist the astrophotographer at virtually all stages of his development.

For the meteor photographer or the piggyback lens photographer, a simple hairdryer or lens heater will prove a most valuable accessory. Unless your camera lens comes complete with a long dew cap, it will only take a few minutes to dew up on a typical damp evening. A hairdryer will quickly solve the problem, provided a mains supply is available. A better solution is a lens heater made from thin, high-resistance wire threaded through a length of hook-and-loop ('Velcro') strip. The strip can be wrapped around the outside of the lens and joined where its two ends meet. By choosing a suitable length of wire, typically a couple of metres, and a suitable battery (I would recommend a rechargeable 12-volt battery with at least 2 amp-hours' capacity), an effective lens heater can be constructed. As a guide, the wire needs to get just perceptibly warm when connected to the battery to be effective; a few degrees Celsius of warming is more than enough to keep dew at bay.

Developing and Printing

Most amateur astrophotographers will soon lose patience with commercial processors of their black-and-white films. Commercial black-and-white processing is expensive, especially for decent-sized enlargements. In addition, the printers will almost certainly have no knowledge of astronomy and are unlikely to centre the image correctly or enlarge the region of interest. In addition, the commercial printers will use a standard grade of paper (astronomers usually opt for high grades, particularly for deep-sky shots) and will charge enormous fees for custom printing (burning in and dodging, etc.). The prolific astrophotographer will soon want his own darkroom equipment.

One of the cheapest, yet invaluable, pieces of equipment is the Paterson developing tank and its equivalents. This is simply a light-proof cylinder containing a plastic or (preferably) metal spiral and a click-tight funnel. In the dark, the film is loaded on to the spiral by alternately twisting opposite sides of the spiral. The spiral is placed in the tank and the lid put on. After this, the whole of the film processing takes place in the light, the developer, fixer and various rinses simply being poured into and out of the tank at the requisite intervals. This simple and inexpensive item immediately frees the amateur from the clutches of commercial film developers. Even if you don't plan to make your own black-and-white prints, I would strongly recommend processing your own films so you can simply select the best negatives for printing.

If you decide to go the whole way and print your films as well (strongly recommended), there are plenty of inexpensive enlargers to choose from, or even darkroom kits containing everything for developing and printing your own films. After a few weeks of experimenting you will wonder why you ever let someone else do your black-and-white printing.

Another invaluable accessory is the *film changing bag*. This is a light-tight bag which enables the film to be loaded into the tank in a fully-lit room without recourse to a cupboard, wardrobe or darkroom. After a tiring night of astronomy it is tempting to go straight to bed and think about film processing the next day. However, loading last night's film the next day can prove difficult if there is no darkroom available. A film changing bag can be particularly useful in these circumstances, and the author has found them invaluable for loading sunspot films in the daytime.

One word of warning, though; in hot weather the user's hands can become very hot and sweaty inside a sealed changing bag, and films can be ruined by contact with sweaty fingers in the cramped space within the bag. Technical gloves, such as VT handling gloves, will solve this problem.

The Telescope–Camera Interface

For photography through the telescope (as opposed to piggyback photography with a camera lens) some method of interfacing camera to telescope is required. Most telescope suppliers stock a range of camera adapters to join most camera bodies (bayonet or screw thread) to 1.25-inch (31.7-mm) or 2-inch (50.8-mm) draw tubes. Some of these adapters allow an additional cylinder to be inserted between draw tube and camera to enable eyepiece projection for lunar and planetary photography.

For the deep-sky or comet photographer the most important features to bear in mind are the accessibility of the focus and the vignetting of the light cone. For optimum performance at the Newtonian focus, a low-profile helical or electric *moto-focus* focuser is recommended. These are much smoother to operate than the standard rack-and-pinion types. (For the CCD camera user a moto-focus has the added benefit of enabling focusing of the telescope to take place while the observer is looking at the monitor screen.)

Of course, most Newtonian reflectors come supplied with some form of focuser; however, these are rarely suitable for astrophotography, and the first step an astrophotographer should make is to install a decent model.

A low-profile focuser is important because it enables the camera to get as close as possible (on a Newtonian) to the secondary mirror (Figure 6.1). As well as increasing the probability of camera focus being achieved at all, this increases the area of the unvignetted field of view. Vignetting is one of the most irritating problems in deep-sky photography and can easily ruin an otherwise excellent photo. To minimise the degree of vignetting the amateur needs to buy a low-profile focuser with a draw tube of *at least* 2 inches in diameter.

If photography is your main aim, a standard 1.25-inch focuser will be totally useless and may well typically *halve* the light grasp away from the centre of the negative. (Incidentally, the camera body itself will cause some vignetting unless your telescope is slower than about *f*/6, in which case the effects will be negligible.)

If you intend doing a lot of visual work as well as photography, then you may have to compromise in your choice of focuser, i.e. low-profile or not, for a

Figure 6.1 A low-profile focuser, to reduce vignetting as far as possible. Note the 3× magnifier attached to the camera's viewfinder, which helps to make sharp focusing more easy to achieve.

Newtonian. However, even if visual work is your main interest, then a 2-inch focuser must be regarded as essential these days, if only to exploit the range of superb wide-field eyepieces available.

The position of the Newtonian draw tube varies considerably from the photographic to the visual position, and the option of inserting filters, off-axis guiders and other auxiliary equipment has to be taken into consideration.

My advice to the potential purchaser of a focuser is to consider the present and future applications of the telescope carefully before a decision is made. If the telescope under consideration is yet to be built, the choice of focuser may affect the size of the secondary mirror, if a Newtonian reflector is used.

To determine the optimum focuser and secondary mirror size for your system, a simple ray diagram, showing the light cone, needs to be drawn. If prime-focus photography is contemplated (the option should always be considered, even by the hardened visual observer), then the focal position should be at least 65 mm out from the mouth of the focuser when fully in, i.e. flat against the tube. If an off-axis guider is to be employed then 80 mm is a better distance.

Unfortunately, the calculations do not end there; other accessories may be positioned in the light path before the camera, necessitating a point of focus even further out from the telescope tube! Filters and coma

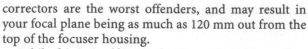

correctors are the worst offenders, and may result in your focal plane being as much as 120 mm out from the top of the focuser housing.

While this in itself is not detrimental, problems then arise for the visual observer. Most eyepieces focus when the focal plane is just outside, or just inside, the field lens. Thus, in a system like the one just described, the rack mount/focuser tube may need to extend as much as 150 mm to hold the eyepiece in position! Not only does this result in an unwieldy system, it necessitates a draw-tube of similar length, which may vignette the light cone in fast optical systems. For example, to illuminate fully a mere 20-mm circle on film at f/5 requires the throat of the draw-tube 150 mm away to be approximately $(150/5) + 20 = 50$ mm in diameter. This presents a very borderline situation, even with a 50-mm diameter draw-tube, to avoid vignetted images.

An allied problem the author has encountered is that some helical focusers described as 2-inch are only this wide at the hole into which the draw-tube fits. This is OK for photographic and visual use, with appropriate 'T' and 31.7 mm adapters. However, it is totally useless for using 2-inch diameter wide-field eyepieces. If you have a focuser of this type then you will need to get a friend with a lathe to make you a 2-inch 'external' to 2 inch 'internal' adapter if you are contemplating purchasing a 2-inch eyepiece.

Many of these problems are irrelevant where Schmidt–Cassegrains and Schmidt–Maksutovs are concerned, for two reasons. First, the prime manufacturers of these telescopes have discovered all the problems, and they provide accessories for most eventualities. Second, these telescopes focus by moving the primary mirror, and the subsequent change in focus point is considerable, alleviating any problems of access to the focus. The accessories stay put while the mirror moves. This said, though, Newtonians are far easier for the 'do it yourself' telescope builder to construct.

It can be seen from the above that the choice of focuser is a tricky one, particularly for the astrophotographer who may want to observe visually. If you cater for every accessory known to man (off-axis guiders, filters, coma correctors, etc.) then your draw-tube will be too long for comfort in the visual position, causing potential vignetting problems and necessitating a larger secondary mirror, if a Newtonian is being used. The advice given in the next section may help you reach a decision.

Guiding, Focusing and Related Considerations

An off-axis guider (an additional small Newtonian secondary and high-magnification eyepiece using the primarily as its objective) is not an essential accessory, and it takes up space in the light path.

A rigidly mounted separate guide telescope will allow you to guide a Newtonian for 10 minutes or so without excessive flexure between guide scope and telescope. Ten minutes will take most Newtonian photographers to their sky-fog limit anyway.

If you really want to go for hour-long photographs then you would be better off with an off-axis guider on a Schmidt–Cassegrain. Apart from the fact that commercial off-axis guiders are totally compatible with Schmidt–Cassegrains, the slower optical system (typically $f/10$) lends itself to long exposures; i.e. sky fogging will be greatly delayed with the slower optical system.

The main advantage of this kind of guider is that by diverting a small portion of the light near the focus to a separate guiding eyepiece it allows the photographer to guide on a star near to the film plane. This almost guarantees that telescope flexure will not affect the final result. The system is ideally suited for Schmidt–Cassegrains; the compact tube assemblies mean that the guiding eyepiece is usually at a convenient height.

The main disadvantages to be found with off-axis guiders are as follows:

- Unless a coma/field corrector is used, the star (by definition off-axis) will be considerably distorted in the guiding eyepiece. They can be almost impossible to use at the focus of a large Newtonian, which may be seven to nine feet off the ground. They do not readily lend themselves to comet photography unless highly specialised versions (on-axis guiders) are used to guide on the nucleus of a bright comet.
- If you use a medium to large Newtonian you will soon realise that a separate guide telescope is a far more civilised way of guiding a deep photograph, unless you want to take exposures of longer than 10 to 15 minutes or so, in which case flexure between tube and guide telescope may well cause problems.
- If comet photography is envisaged then a separate guide telescope will be a virtual necessity.

This third point opens up another subject, namely, how does one guide on a comet? If there is one area of astrophotography that has not been commercially exploited, this must be it! Comets move relative to the background sky and therefore need to be tracked by moving the telescope in a predetermined direction at a predetermined rate. A number of computer programs are available for determining the direction and rate of a comet's motion in arc seconds per hour (3600 arc seconds = one degree of arc).

Once the required direction and rate of motion are known, the problem is how to achieve them. In days gone by, the professional astronomer (and the advanced amateur) used *bifilar micrometers* in guide telescopes. At regular intervals the micrometer webs would be moved and the guide star recentred, thus moving the telescope in tiny increments away from the guide star. Bifilar micrometers are very expensive, and so amateurs have sought other routes to enable them to track comets. One can easily obtain illuminated reticule eyepieces with graduated scales (the Celestron Microguide eyepiece is an excellent example), and then simply use a PA scale glued to the guide telescope draw-tube to set the PA (Figure 6.2). If you centre a star at zero degrees declination in your guide telescope and switch the drive off then the star will drift at 15 arc seconds for every second of time. This will enable you to calibrate the graduated scale in arc seconds per division. (You may need a Barlow lens to extend the focal length of your guide telescope, if the reticule divisions are larger than 5 arc seconds.) By turning the eyepiece so that the star drifts along the graduated scale with the drive off, you can calibrate the East–West line. To guide on a comet you just need to turn your eyepiece through the required angle and calculate how often you need to reposition the guide star along the scale. (Remember that your guide telescope inverts the field!)

Personally, I prefer to offset the guide star without glancing at my watch. A cassette tape recorder can prove invaluable in this regard. By prerecording audio messages at the required time interval, e.g. 'move to division two ... move to division three', you can concentrate fully on guiding, without distraction.

Some amateurs have built automatic comet trackers which move the camera, or the guiding eyepiece, at the right rate and position angle to track a comet; if you have a workshop and a good knowledge of electronics then this is definitely the way to go.

Figure 6.2 Comet tracking. A PA scale glued to the drawtube of the guide telescope provides a crude, but effective and inexpensive, alternative to the bifilar micrometer.

Coma correctors also take up space in the light path and can introduce vignetting. Large Newtonians of $f/5$ or slower will benefit least from their use, because the field of view will be too small to introduce coma, except at the edges. As most astrophotographers tend to enlarge their prime-focus images by about 10×, properly centred objects will not suffer adversely from coma. So a coma corrector is not essential unless you are using a telescope below 250 mm aperture and faster than $f/5$, or if you plan to use medium-format (i.e. larger than 35-mm) film. If you are building your own telescope of this sort of aperture then $f/6$ or $f/7$ will give you definite advantages where coma and vignetting are concerned.

In summary, if you can do without a coma corrector and an off-axis guider then you can have a very compact focusing system; with the Newtonian draw tube fully in you're at the prime focus photography position, and 80 mm further out you can accommodate most eyepieces.

Helical and motorised focusers are strongly recommended for astrophotography and CCD imaging, respectively, where focusing needs to be accurate and easy.

If you have a limited budget then the bottom-line advice is: *don't* use a 1.25-inch focuser/rack mount for astrophotography; make sure you can reach the prime focus with a camera attached and the focuser fully in.

On the subject of accurate focusing I would also strongly recommend two other accessories. The first of these is a 3× magnifier for the SLR viewfinder; this will enable a sharp focus to be achieved on star images for deep-sky work (more advanced 'sure-sharp' focusers are also available from astro-specialists). Second, a very clear ground-glass screen will prove invaluable for astrophotography, particularly for lunar and planetary work at long focal lengths. Most SLR manufacturers can supply interchangeable screens.

Film Hypering Kits

In the last fifteen years more and more amateurs, particularly in the USA, have been hypersensitising ('hypering') film to improve its performance.

The hypering process removes moisture from the film and lightly pre-fogs the emulsion to form sites around which grains can form. This virtually eliminates the usual film reciprocity failure mechanism which prevents very faint objects from ever being recorded (there is a minimum level of light below which the effective film speed falls sharply). Hypering is achieved by vacuum-baking the film in forming gas or hydrogen.

Hypersensitising fine-grain films like Kodak 2415 Technical-Pan results in an emulsion that can continue to gain faint stars long into the exposure and the inherently slow, fine-grain emulsions have a superb 'signal-to-noise ratio', enhancing the detection of faint objects.

The only way of eliminating reciprocity failure before gas-hypering came along was by using a cold camera in which the emulsion was cooled to around –50 °Celsius.

Hypered film can be supplied by various companies in the USA and UK, or the amateur can purchase a Lumicon gas-hypering kit and hyper his or her own film. Full hypering instructions for black-and-white or colour film are supplied with each kit.

Hypering is not essential, but if you want to take pictures without excessive grain, and if you want to go fainter than about mag. 17 then it is highly recommended. Hypered Kodak 2415 film will also enable astrophotographers in urban locations to get deeper than they could using fast standard film, as the slower film is much more resistant to fogging.

CCDs, Autoguiders and Image Intensifiers

So far, I have avoided mention of autoguiders. The SBIG ST-4 is currently the best known autoguider on the market. It is essentially a small CCD chip with software which enables it to monitor the drifting of a star in the focal plane. When the star drifts, appropriate north, south, east or west pulses are sent to the drive motors to recentre the star. The ST-4 replaces the human being at the guiding eyepiece. Obviously, some compatibility is required between the ST-4 and the motors driving the telescope. As with off-axis guiders the market is heavily biased towards owners of Celestron or Meade Schmidt–Cassegrains, although some other telescopes made in the USA are catered for.

My advice about autoguiders is similar to that for off-axis guiders: if you can afford one (and they can be used for CCD imaging as well) and you intend doing long exposures with a Schmidt–Cassegrain then they will save considerable pain and suffering at the guiding eye-piece; however, they are far from inexpensive.

On the other hand, if you have a large Newtonian and don't intend to expose beyond 10 to 15 minutes, then the setup time for the autoguider will not justify its use. Autoguiders come into their own when 1-, 2- or 3-hour exposures are being contemplated.

I've often thought that if you are using a CCD auto-guider for hour-long exposures then you might just as well put the CCD at the focus and bring the exposure down to minutes anyway! This is only partly true, of course, as the wider field and unpixelated quality of a good astrophoto will always give a more pleasing pic-ture (at least until larger CCD chips with smaller pixels come along).

CCD cameras warrant an entire book to themselves. They are expensive accessories beyond the reach of many amateurs and – in my opinion – have in no way replaced photographic film. Their main advantages are that they have about a two-magnitude advantage over even hypersensitised film, and that the digital image lends itself to powerful image-processing routines which can bring out hidden detail.

The results of image processing on amateur CCD images of the planets can be quite astounding. There is also the fact that image processing allows the urban observer to process out the light pollution as if one were observing from a country site. Like many (but not all) things that are good, image-processing software tends to be costly.

If you want to end up with photographic prints, pho-tographs of the monitor screen rarely look as æstheti-cally pleasing as those taken on fine-grain film. So if you can't afford a CCD camera don't feel left out; film is still good stuff!

As well as CCDs another device has (re)entered the fray of late, namely the *image intensifier* (vacuum-state image intensifiers were the earliest kind of 'light amplifiers' used by astronomers). I have found that there is some confusion between CCDs and image intensifiers. The difference is this: a CCD chip collects photons, as electrons, in each pixel for the duration of an exposure, which may be a tenth of a second or even 10 minutes. The accumulated charge held in each pixel is downloaded at the end of the exposure and subsequently displayed on a monitor. An image intensifier, on the other hand, doesn't 'collect' photons at all. As each photon arrives it is converted into an electron, amplified as much as 100 000 times and immediately hurled onto a glowing phosphor screen. The amplification factor is impressive, but the limiting magnitude is not, and is largely governed by the persistence of the phosphor screen. The main rôle of an image intensifier (when suitably coupled to a TV camera) is for filming dim events in real time, for example grazing lunar occultations, asteroidal occultations and meteor showers.

Computerised Observation

Electronic Gadgets

The electronic revolution and the increasing complexity of microprocessors/microcontrollers have spawned a whole range of 'toys' for the amateur astronomer. The fully automated observatory is almost here, at a price! Computer control, digital setting circles, e-mail and CD-ROMs would have been pure science fiction twenty years ago, but are now almost commonplace.

But how much of this paraphernalia does the amateur really need?

Strictly speaking, none of it. As an advanced amateur myself, with numerous gadgets and toys, I often feel that life would be a lot simpler with the absolute minimum of equipment! One only has to look at the discoveries of George Alcock – he used nothing but binoculars – to see that one can enjoy astronomy without investing in huge amounts of equipment. Many 'electronic-age' accessories might be regarded as luxuries, to indulge in if finances allow.

One of the most common and popular electronic accessories is the 'digital setting circle'. Various models are available and, as usual, the Schmidt–Cassegrain user is better catered for. Having said this, retrofitted units are available from specialists for most types of mounting, including Dobsonians.

A device of this sort uses a *shaft encoder* on the polar and declination axes. Pulses from the motion of the shaft encoder tell the telescope how far it has moved since it was initialised (on a known star) and the corresponding right ascension and declination are shown on an LED (light-emitting diode) display. By the way, do bear in mind that the accuracy of standard digital setting circles is no better (and often worse) than that of mechanical circles; 'electronic' isn't necessarily the same as 'more accurate'.

If you are an astrophotographer who uses circles blindly to position the telescope (i.e. without checking the field visually) then you may well be better off with 'old-fashioned' mechanical circles. For example, many digital circles show the RA to the nearest minute of arc. This is a quarter of a degree on the celestial equator and can make all the difference between an object being in the sharp centre of the field and not. However, if you are a visual observer, digital setting circles can be great fun, particularly as they come with a digital library of deep-sky objects, enabling the observer to find and/or identify dozens of objects per night. A visual supernova hunter would find such an accessory ideal.

Going one step further, one can now buy (from Celestron or Meade) computer-controlled telescopes that will actually slew one to the desired object in seconds. These telescopes are fascinating to watch as they whirr and whizz to the next object. One word of caution though: the motors make quite a noise when fast-slewing, perhaps enough to wake the neighbours at 3 am. Nevertheless, and despite their high cost, automated telescopes are certainly great fun!

Reviews in the American magazine *Sky & Telescope* are the best guide to the performance of most accessories and, in many ways, a subscription to this magazine is probably the best astronomical investment you can make if you are contemplating spending hundreds or thousands of pounds on equipment.

Computer Software

As well as electronic hardware, the electronics revolution has spawned faster and faster personal computers (PCs) and increasingly sophisticated software. There is no other area of affordable technology where advances have been so rapid. The calculating speed of a modern PC has increased nearly a thousand-fold in the last five years, and hard disk sizes have become enormous. You can now purchase a set of two compact discs (CD-ROMs) which give the RA, declination, and magnitude of *19 million* objects in the *Hubble Guide Star Catalog* for a price of around $70 (Figure 6.3). Software is readily available for displaying the Hubble data in the form of star charts; other sophisticated programs can take you on a simulated tour of the Solar System, the Galaxy or the Universe.

Much of the available software is purely for entertainment and education, but some of it can undoubtedly help the advanced amateur in the location of faint objects. However, this is a clear case of 'buyer beware', for a lot of the expensive commercial software does not actually deliver what the serious observer requires. For example, one of the first requirements of the amateur astronomer might be to find out where, precisely, that new comet is in the sky. All that is needed is a program that allows the inputting of the orbital elements and the outputting of the RA and declination, with the comet's speed and direction as an additional feature. Make sure the software delivers what you want – flashy graphics and multiple windows are no help when fundamental facilities are lacking.

Software such as this is best obtained from your national amateur astronomical association. For example, in the UK, Nick James of *The Astronomer* magazine has produced a suite of programs that is so successful even the professionals have ordered copies. These particular programs do exactly what the advanced amateur wants, no more and no less. Software currently (1995) available from this source includes a highly accurate comet and asteroid ephemeris program; an altitude–azimuth program for plotting a comet's position above the horizon by reading the ephemeris data; highly advanced programs for determining a comet's orbit from a few measured positions and for checking the accuracy of measurements to an existing orbit; an asteroid check program which tells the user whether any of 5200 asteroids are within a specified area (useful for

Figure 6.3 The *Hubble Guide Star Catalog* CD-ROM. The monitor shows a 2° field from the GSC, plotted with software written by Nick James of *The Astronomer*.

supernova suspects); software for enabling star charts to be prepared from the *Hubble Guide Star Catalog* (the charts are coded to enable them to be e-mailed to observers world-wide). In addition, charts of over 100 supernovæ, novæ and dwarf novæ are available on floppy disk.

In addition, the British Astronomical Association Computing Section can supply inexpensive software for calculating lunar occultations, artificial satellite transits and comet ephemerides.

E-Mail

Having mentioned e-mail (electronic mail), it may be worth expanding this point. When a new object is discovered, advanced amateurs will want to hear the news as soon as possible. If conventional postal services are used to convey the news, it may be several days before the observer is alerted. An amateur astronomer can receive rapid news of new discoveries by acquiring a PC and a modem and by subscribing to an e-mail service. Various services are available to choose from, the major systems being Compuserve and Internet. E-mail services are now inexpensive for the basic service, but some of the extended services cost more. To use e-mail you need a modem, or a modem card, and – of course –

a suitable computer. Software isn't a problem; for example, Compuserve provide their information management system, WinCIM, either free or for just a few pounds or dollars.

An alternative approach is to log on to a voluntarily operated bulletin-board system or simply buy a fax machine. These options are cheaper but exclude the possibility of exchanging confidential mail messages with friends on the same system.

Discovery Equipment

Sooner or later, every amateur starts to wonder if he could make a discovery. Amateurs have a long history of comet and nova discovery and, in the last fifteen years, supernovae have been added to the list. It should be stressed that luck rarely plays a part in the discovery game. If you don't have endless patience and are not prepared to spend hundreds of hours a year searching then forget it. But if the determination is there, what equipment will enhance the chance of success? There are three main categories of object which are hunted by the amateur: comets, novæ and supernovæ.

Comets are probably the simplest objects to hunt for; they are fuzzy in appearance and so easily stand out against the background stars. Apart from a suitable telescope and a good star atlas (showing all the nebulæ and galaxies) no auxiliary equipment is needed. As far as the instrument is concerned, it should be easy to use and have a wide-field eyepiece. The world's most successful hunters use a variety of magnifications and apertures. 25 × 150-mm binoculars are quite popular among the Japanese, but David Levy uses a 400-mm reflector with a magnification of about ×50.

For novæ and supernovæ, the patroller may be using visual or photographic means. The visual nova patroller simply knows the sky so well that he has memorised the constellations (to magnitude 8 in the case of George Alcock!); in this case a star atlas to confirm the discovery is all that is required. Some visual supernova hunters (for example, Robert Evans) use a telescope instead of binoculars and may have hundreds of galaxy charts and photographs at arms length, but no other equipment is required. (For serious supernova hunters the supernova search charts by G. D. Thompson and J. T. Bryan are highly recommended (Figure 6.4).)

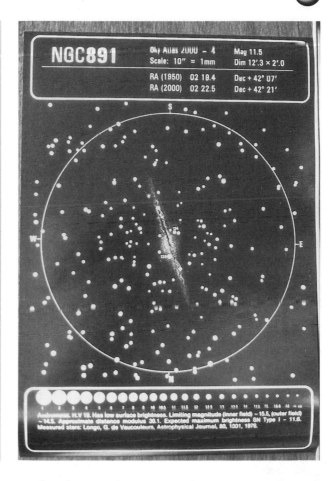

Figure 6.4 The galaxy NGC 891, shown on one of the excellent supernova search charts produced by G. D. Thompson and J. T. Bryant.

The photographic nova or supernova patroller will need some way of comparing last night's negatives with his master reference set. There are basically two ways of doing this: with a *stereo comparator* or with a *blink comparator*. A stereo comparator simply consists of two identical magnifiers (typically giving 10× magnification), one for each eye, which are positioned over each negative. The negatives are placed on a light table and held flat either by a slab of clear acrylic or by mounting in slide mounts.

The cheapest magnifiers are the plastic eye-loupes obtainable at any photographic store. Providing the user has good stereo vision, he or she will be able to stereo-merge the two negatives (assuming they are of similar limiting magnitude and cover exactly the same field). Any new object will immediately stand out and give the impression of floating above or below the negative. If a new nova is suspected then a check should first be carried out to determine whether any bright asteroids or variable stars are in the field. (A nova patrol with a 50-mm lens will typically reach magnitude 10.)

A copy of the GCVS (*General Catalogue of Variable Stars*) will be invaluable for checking out the fainter variable stars, and the *Handbook* of the British Astronomical Association (or equivalent national publications) contain ephemerides for the brighter asteroids. If you are only checking to about magnitude 9.5 a copy of the excellent star atlas *Uranometria 2000* (Figure 6.5) will cover all the variable stars you are likely to pick up.

If you are supernova hunting then you will almost certainly be going down to at least magnitude 14, and faint asteroids and variable stars will definitely be a problem. If you have a PC then I would strongly recommend that you purchase a copy of the program called AST.EXE by Nick James of *The Astronomer* magazine. On a PC with a 486 processor this will tell you within about five seconds whether any of the first 5200 asteroids are near your galaxy! Incidentally, the GCVS is now available on a CD-ROM for PCs equipped with the necessary player.

I have so far assumed that the hunter will be able to use a stereo comparator. However, a significant fraction of the population have less-than-perfect (or non-existent) stereo vision and may prefer to use a blink comparator. The best known form of blink comparator for amateur use was invented by Los Angeles amateur Ben Mayer, and is called the Problicom (from projection blink comparator); basically anyone can build one. The Problicom is constructed of two identical 35-mm slide projectors. One is mounted above the other in a housing which enables them to project their slides on to exactly the same point. A semicircular, motor-driven shutter then rotates in front of the two projector lenses, alternately cutting off first one light beam and then the other. The projected image thus blinks between the patrol and reference shots. The Problicom is a neat idea, but it can be very fiddly to adjust between slides. The top projector really needs to have fine adjusting screws on each axis to make the aligning easy.

Other forms of blink comparator have also been devised by amateurs. Denis Buczynski of Conder Brow Observatory in the UK has a very impressive video comparator which alternately blinks two photographic plates on a TV screen; this is definitely the easiest way to do the checking!

Some amateurs use a system in which the patrol and

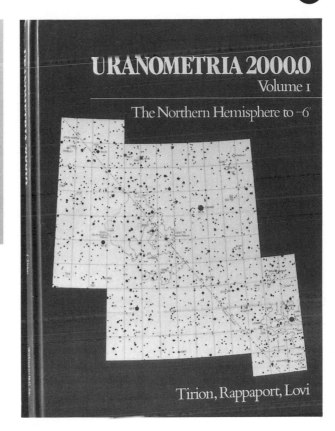

Figure 6.5 Volume 1 of *Uranometria 2000*, an essential aid for nova-hunting to magnitude 9.5.

reference slides are viewed by different eyes (as with the stereo comparator) but blinked alternately. I have never understood the principles behind this technique; indeed, I have never found anyone who recommends it, so how it became so popular is a mystery to me. Personally I recommend the stereo comparator or the blink comparator, rather than the 'alternate eye' comparator.

A good comparator is essential for photographic patrolling but, as already mentioned, the prime requisite for a discovery is determination and patience.

Astrometric Equipment

One of the most useful areas in which the amateur can contribute is the precise measurement of astronomical positions. In particular, precise measurements of a newly discovered comet's position can help to deduce its orbit. Very few professional astronomers are still doing this kind of work, and amateur measurements of

comet positions are becoming more and more important. Anyone hoping to contribute to this field needs to
be an accomplished astrophotographer as a minimum
requirement. A comet's position has to be measured
accurately with reference to sharp star images and high-
quality negatives are a prerequisite.

The piece of equipment for this task is called a measuring engine – essentially a microscope with cross-
hairs, under which an illuminated negative can be
moved in the x and y axes with a minimum precision of
10 μm. A piece of equipment such as this can already be
found in industry, where it is generally known as a 'tool-
maker's microscope'. However, the skilled and/or
resourceful amateur can construct his own measuring
engine without too much trouble. Unfortunately, the
construction of a measuring engine is well beyond the
scope of this chapter, but if you can lay your hands on a
low-power microscope with cross-hairs and a couple of
accurate micrometer barrels, a friend with a workshop
will probably be able to make one for you.

In the last few years the advent of affordable CCD
cameras has added another dimension to astrometric
measurements. If a comet, along with enough reference
stars, can be captured on a CCD image, then an astrometric measurement can be made without any
measuring engine being necessary. Software has already
been written by a number of amateurs to measure star
and comet positions from the CCD image with reference stars from the *Hubble Guide Star Catalog*.

Conventional astrometric measurement necessitates
that you are able to identify a field of reference stars for
the reduction process. If you have a copy of the *Hubble
Guide Star Catalog* on CD-ROM, and software for plotting the field, then this process is easy. However, if you
are working with star charts and a negative it may take
several hours to determine which are the best reference
stars. (Incidentally, for precision photographic astrometry at Epoch 2000 you will need to obtain a copy of the
Positions and Proper Motions Catalogue or *PPM*.) It can
simplify the identification of reference stars if you can
overlay your negatives on a star atlas of roughly the
same scale. For example, the excellent *Atlas Stellarum*
by Hans Vehrenberg has a scale of 1° to 30 mm, which is
the same image scale as the negatives from a telescope
of 1.7 metres focal length. Thus if you are using, say, a
300-mm Newtonian of $f/5$ to $f/6$ you will be able to
match up your star-fields very nicely. (Another advantage of *Atlas Stellarum* is that its limiting magnitude of

14 is very similar to that of a typical astrometric negative.) Similarly, *Uranometria 2000* has an image scale equivalent to the negatives from an instrument with a focal length of 1 metre.

Overlaying negatives on a star atlas in this way can save hours of hassle.

Filters

The use of filters in amateur astronomy has increased dramatically in the last 20 years, owing to two main factors: the increasing encroachment of light pollution and advances in filter fabrication technology. Twenty years ago an amateur with 'a set of filters' was someone who had the right set of Wratten filters for planetary observation. Nowadays there are different filters for every conceivable type of visual and photographic astronomy. I will deal with each category in turn.

Planetary Filters

Much has been written on the use of Wratten filters for visual planetary observing and there is no doubt that different observers see different effects through the same filters. Venus, Mars, Jupiter and Saturn observations are all enhanced by the use of filters and there are four specific Wratten filters which are strongly recommended, namely Wratten No. 25 (red), Wratten No. 15 (yellow), Wratten No. 58 (green) and Wratten No. 47 (blue)

These filters are relatively inexpensive and are easily obtained from any professional photographic supplier. Their use will enable faint markings observed in white light to be enhanced and more easily recorded.

Deep-Sky Filters

The problem of light pollution has spawned a new breed of 'interference' filters to specifically enhance the emissions of nebulæ and suppress the emissions from street lights. Probably the best all-round 'light pollution rejection' filter is the Lumicon Deep-Sky Filter, which blocks most sodium and mercury street-light emissions while letting through the vital emission lines of nebulæ. The filter can be used visually or photographically, and I

have found that the combination of a Lumicon Deep-
Sky Filter and an ultra-fast film like T-Max 3200 is very
powerful for nova patrolling from an urban site. The filter does produce a loss of about a magnitude or so on
photographic star images, but this is a small price to pay
if you want to photograph nebulæ from an urban location. The filter allows spectacular colour photographs of
bright objects such as the Orion Nebula to be captured
from city sites without any of the orange sky glow pol-
luting the image.

Lumicon markets a whole range of other filters, most
of which are extremely specialised and beyond the scope
of this chapter. These more specialised filters are specifically designed to enhance single emission lines of hydrogen or oxygen, so as to enhance particular types of object
such as supernova remnants, etc., or the Swan band emission line in gassy comets. There are, however, two photographic filters which are strongly recommended for photography with telephoto lenses. These are the Hα pass
filters and the minus-violet filters; they are relatively
inexpensive but can produce spectacular photographs.

The Hα pass filter totally blocks all light below
630 nm, i.e. it only lets through red light and above. This
means that virtually all forms of light pollution are
blocked and the Hα emission line (656 nm) is not. The
only problem with using this filter is that it must be used
with gas-hypered 2415 film, which is sensitive to the red
end of the spectrum. In addition, the lens focus should be
set roughly midway between the normal and infra-red
focus positions. The contrast gains achieved with this
lens are dramatic; from a city location you can easily photograph Barnard's Loop in Orion and the complex fila-
ments of gas around the Cygnus area. Even more spectacular results can be achieved using a Schmidt camera.

The minus-violet filter, primarily designed for sharp-
ening colour telephoto images of star fields, does exactly what you would expect: it blocks the violet part of the
spectrum, which gives rise to the bloated star images
seen in many amateur photos. This inexpensive filter
will drastically sharpen colour telephoto photographs
of the night sky, although you will need to increase the
exposure time to achieve the same limiting magnitude.

Solar Filters

I hesitate even to mention solar filters, as looking at the
Sun through any telescope is a highly dangerous pursuit

unless you know precisely what you are doing. Just because the Sun may look dim enough to observe through a filter does not mean that it is. Infra-red radiation cannot be seen by the eye, and one can easily be lulled into a false sense of security by an image that is not dazzling. Many amateurs have looked through filters at a pleasant solar image unaware that large amounts of infra-red energy are pumping into their eye. The result has been PERMANENT BLINDNESS.

Manufacturers STILL advertise and sell solar filters that are unsafe. Indeed, the vast majority of solar filters sold with small refractors are TOTALLY UNSAFE.

My advice here is simple. If you want to observe the Sun visually, project it on to card; if you want to photograph the Sun, contact the Director of the British Astronomical Association for advice. Alternatively, read the excellent article on Solar Observing by Don Trombino in the *1994 Yearbook of Astronomy*.

Photometric Equipment

Photometry, i.e. the precise measurement of magnitudes by using electronic detectors, is a highly specialised branch of astronomy using highly specialised equipment. A guide to photometry is well beyond the scope of this chapter, but essentially one needs a photomultiplier tube and a precise set of calibrated filters for each waveband being considered. This sort of field is so specialised that one must seek expert advice from leading amateurs on a national level. Photometry enables amateur astronomers to contribute to professional observing programs where accuracies of 0.01 magnitude are required. Thus an inaccurate measurement in this context is worse than no measurement at all.

CCD cameras are now being used by some amateurs to measure magnitudes, but extreme caution is advised here. The filter sets used by amateurs with photomultiplier tubes have been proven over decades to be reliable. CCD photometry is in its infancy, and it will be some years before the right filters are determined for the CCD cameras in amateur hands. Even then the results will be treated with some caution until the reputation of the amateur and CCD photometry is established.

Equipment Suppliers

All of the following distributors and suppliers are highly recommended.

For frank and detailed reviews of astronomical accessories, take out a subscription to Sky & Telescope.
- *Sky & Telescope*, PO Box 9111, Belmont MA 02178-9111, USA; telephone (800) 253-0245.

For UK purchases of American-made equipment, the largest British distributor and agent for the leading US companies is
- Broadhurst Clarkson and Fuller, Telescope House, 63 Farringdon Road, London EC1M 3JB, UK.

The UK distributor for Celestron telescopes and accessories is:
- David Hinds Ltd; telephone (UK) Tring (01442) 827768.

For world-wide mail order of deep-sky filters and hypering kits, contact:
- Lumicon, 2111 Research Drive #5S, Livermore, CA 94550, USA; telephone (510) 447-9570.

For digital setting circles, motorised and quality focusers and other retro-fit accessories, contact:
- JMI, 810 Quail St., Unit E, Lakewood, CO 80215, USA; telephone (303) 233-5353.

For CCD autoguiders and cameras, contact:
- SBIG (Santa Barbara Instrument Guide), 1482 East Valley Road, Suite #J601, Santa Barbara, CA 93108, USA; telephone (805) 969-1851

For the British-made 'Starlite Xpress' CCD Camera, contact:
- FDE Ltd; telephone (UK) (01734) 342600.

For inexpensive software for comet and asteroid ephemerides, and Hubble Guide Star CD-ROM chart creation, contact:
- Guy Hurst, *The Astronomer*, 16 Westminster Close, Kempshott Rise, Basingstoke, Hampshire RG22 4PP, UK.

Atlas Stellarum *is obtainable from:*
- Treugesell-Verlag, Dr Vehrenberg KG, Schillerstraße 17, D-4000 Düsseldorf 1, Deutschland.

For a subscription to e-mail services, contact Compu-

serve at one of the following addresses:

- Compuserve, 1 Redcliff Street, PO Box 676, Bristol BS99 1YN (UK)
- Compuserve, Postfach 1169, D-82001 Unterhaching bei Munchen (Germany)
- Compuserve, 5000 Arlington Centre Blvd, PO Box 20212, Columbus, OH 43220 (USA).

For star charts, catalogues and CD-ROMs, contact Sky & Telescope *at Sky Publishing or Broadhurst Clarkson at the addresses given above.*

Chapter 7
Electronics and the Amateur Astronomer

Maurice Gavin

Computing

Many amateur astronomers own or have access to personal computers. They may be used for anything from word processing and report writing, through computing the precise position of a comet or asteroid, to recording and analysing images captured by CCD (charge-coupled device) cameras. Neither a telescope nor a CCD camera is necessary to do image processing. Images from space agencies and experienced amateurs are available on disk to try one's skills upon with suitable software.

Increasingly sophisticated programs are available on disk to convert the PC into a planetarium, or to allow the user a 'flight of fancy' in stunning graphics through the solar system and beyond. Today many of the pictures returned to Earth from interplanetary craft are available in floppy-disk or CD-ROM formats. Figure 7.1 demonstrates what can be achieved by the amateur PC user. Amateurs even have access to the immense star catalogues like the *Hubble Guide Star Catalogue* (*GSC*) used for pointing the Hubble Space Telescope. This is limited to stars down to magnitude 15 to 17 – a level that can be recorded in just a few seconds with an amateur CCD camera. CCD images down to mag 20 and fainter are possible from amateur equipment. Dedicated amateurs will thus expect even 'deeper' catalogues on disk, perhaps initially down to the mag 20/21 limit of the

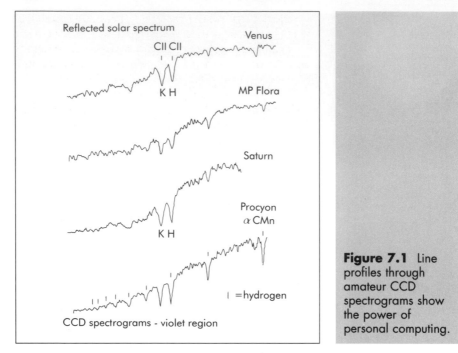

Figure 7.1 Line profiles through amateur CCD spectrograms show the power of personal computing.

Palomar Schmidt camera plates which cover the entire sky. It is not beyond the bounds of possibility that one day amateurs will have direct access via their home computers to orbiting telescopes to point, record and download images of their own choice. Clearly a specific programme of study would be necessary for such a link-up beyond astro-sightseeing.

Image Intensifiers

Image intensifiers are popularly called night-sights. With their military and security applications, access to them used to be strictly limited, but they are now commonly available (if expensive) in most countries. Modern image intensifiers are portable electro-optical devices (EODs) that effectively boost the light output across two glass plates by a factor of several hundred times or more by using accelerated electrons. The plates are made of fibre-optic bundles and were among the first devices to use this technology. Earlier versions (first generation) are typically 70 mm in length by 50 mm in diameter, with the viewing screen (output plate) about 20 mm to 35 mm in diameter. They may be linked in a cascade (typically no more than three units) to boost the signal further. Second generation devices are much slimmer (typically 10 mm) and use a technique of microchannelling to accelerate the electrons through fine glass tubes thinner than a human hair. Figure 7.2 shows a three-stage image intensifier.

The image intensifier can be inserted into a telescope in place of the eyepiece with the image focused on the input faceplate. The viewing screen (usually bright-green phosphor) is viewed with a low-power simple lens of about 50 mm focal length. Alternatively the screen can be photographed with a camera placed a few centimetres behind the viewing plate.

Image intensifiers have a number of drawbacks. The picture displayed is fairly coarse and 'sparkly' because of electronic noise. Some first-generation image intensifiers cause severe distortion of the image, particularly around the edge of the field. The image tends to be of low contrast, which negates the usefulness of its high brightness. Simple geometry also indicates that most of the light exiting from the rear display plate escapes the eye (or camera lens) altogether, so that the overall efficiency of the system is low. It can be much improved if the recording film is placed in physical contact with the output screen. Modifications to the image intensifier circuitry may be possible (and necessary) to reverse the polarity of the system and to make the viewing plate of zero voltage in the latter mode. In some professional applications the viewing plate is bonded direct to a CCD detector surface for the best sensitivity for spectrographic cameras. Such units have yet to filter down to

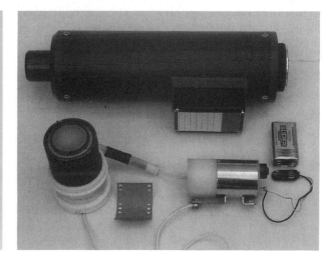

Figure 7.2 A three-stage image intensifier (eyepiece on the left). The film is held in contact with the faceplate for maximum quantum efficiency.

the level of the amateur market, but may be available before too long.

Notwithstanding the above, image intensifiers will typically show stars visually to one to two magnitudes fainter than can be seen without the device. Because their peak spectral sensitivity tends towards the red and near infra-red region, they have applications for the examination of red stars (Mira variables, etc.) near minimum light when these objects cannot be seen directly. Amateurs have coupled image intensifiers to video cameras to record the night sky in real time for a meteor watch showing wide areas of the sky nearly to naked-eye limits.

Sometimes image intensifiers are available on the military surplus market and are suitable (and cheap enough) for experimentation. Although run from a small torch battery they generate an extremely high voltage (but very little current); the units must be properly insulated for safety and handled with care. Some electronic skill is essential.

TV Security Cameras

TV security cameras may also be obtained on the surplus market. They may have modest picture quality but make up for it in sensitivity, especially in the near infrared. They are ideal for linking to a telescope if the requisite skills are to hand. They are generally limited in exposure time to a TV frame of about a twenty-fifth of a second, and this is too brief to record faint objects like nebulae and galaxies. The Moon and planets make good targets. Images can be captured on a video recorder or frame-grabber for manipulation through a computer with suitable software. The brief exposures ensure that moments of good seeing can be utilised.

Home Video Cameras

Many astronomers own or have access to a camcorder – a combined miniature video camera and recorder. Modern camcorders are sensitive enough to target the Moon and all the brighter planets from Mercury through to Saturn and in colour. They are an excellent medium for presenting astronomy to a lay audience, friends and societies. Because all camcorders (except the semiprofessional versions) have fixed lenses that are not removable, the camera must be held at the telescope eyepiece in *afocal mode*. Both eyepiece and camera lens should be focused to infinity, and the camera autofocus setting switched off. Because of this juxtaposition the camera lens should be zoomed to a telephoto setting to avoid vignetting around the edge of the picture. Use an eyepiece of about 25 mm focal length initially and a screw-in UV filter over the camera lens for protection. Alternatively the camera can be used without additional optical aid to image sky phenomena from a fixed tripod or small equatorial platform. If a time-lapse mode is available then an eclipse of the Moon lasting several hours can be reduced to a few sprightly minutes, thus animating on playback motion that is otherwise imperceptible.

There may be no control over exposure or colour balance. However, modern cameras use CCDs and are extremely sophisticated; they will automatically boost the signal as required. Some even have electronic circuitry to stabilise a jumpy image caused by camera shake! If the image is clear in the electronic viewfinder then it can usually be recorded. Reducing the colour saturation of the picture on playback, by turning the TV colour control knob to minimum, will reduce or eliminate picture 'noise'. For any serious work the camera should be supported on a platform attached to the telescope and correctly balanced. However, a camera handheld to the eyepiece will give very pleasing results. Sound recorded with the picture, perhaps including a commentary, greatly adds to the ambience on playback. Use the camera's date-and-time facility for subsequent identification of the pictures.

Events such as a bright stellar occultation by a planet can produce science when the video recording is analysed.

Computer-Aided Telescope (CAT)

This usually takes the form of a small computer that plugs into the drive system of the telescope to aid pointing at faint and difficult targets. Some units contain a whole library of thousands of objects, including planets, asteroids, and deep-sky nebulæ, star clusters and

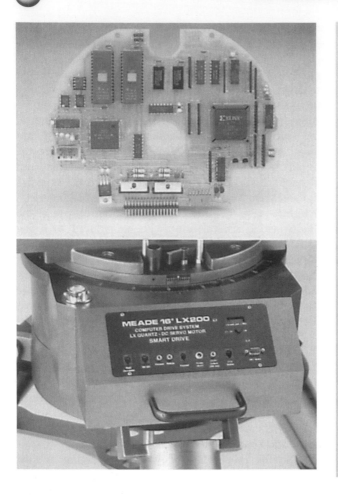

Figure 7.3 Circuit board (above) and housing (below) for the Meade *LX200* series of CAT which, mounted beneath the RA drive, will locate and point the telescope towards any planet or any one of 64,000 deep-sky objects, working in either altazimuth or equatorial mode.

Picture courtesy of BCF Ltd.

galaxies. The CAT shown in Figure 7.3 can locate and point the telescope at any one of 64,000 deep-sky objects. Such devices allow the user to find many more targets in a evening than is possible through star hopping from a star chart. They are particularly useful for newcomers to the sky. It is possible to link a database on a PC direct to the telescope and use it to present the view visible through the eyepiece or camera. This makes the identification of a new star (nova or supernova), asteroid or comet more certain by its absence from the catalogue image. Some CATs will work on altazimuth telescopes like the Dobsonian or larger SCT, mounted this way to minimise the engineering problems common to equatorial mounts. The latter may have motorised prisms before the eyepiece or camera to counteract field rotation in altazimuth mounts.

Two shaft encoders (one on each axis of the telescope) and a 'fix' on two bright stars is sufficient for the computer to 'visualise' the whole sky in its memory. Some CATs, which contain a complete calendar/clock in memory, can do this from a one-star fix. This sounds good, but it assumes an altazimuth telescope that is mounted precisely level (no mean task) or an equatorial mount that is precisely aligned on the celestial pole. The two-star option is easier and potentially more accurate.

Periodic Error Correction (PEC)

Few amateur telescopes track the stars perfectly, even when correctly aligned on the celestial pole. Such tracking errors may blur long-exposure photography through the telescope without constant manual correction via a guide eyepiece. PEC is a computer-controlled device that memorises the tracking errors of the drive system (usually on one cycle of the worm drive) and changes the motor speed constantly throughout the cycle to compensate. This greatly reduces the frequency of applying manual correction. PECs are typically

Figure 7.4
Meade's *Smart Drive* PEC can be 'trained' to recognise and correct periodic mechanical errors in the telescope's drive system to within 5 seconds of arc.

Picture courtesy of BCF Ltd.

integrated into the telescope drive system and come as part of the package. Some PECs are factory-preset. Others, such as the one shown in Figure 7.4, can be 'trained' by the user. Tracking errors can be reduced by a factor of up to ten times to errors as small as 3" to 5" of arc per minute at best in commercial telescopes.

PECs are of no particular value to visual observers, who don't need this level of tracking accuracy.

Electronic Star Guider (ESG)

Although PECs reduce tracking errors, they don't eliminate them completely. Also, there are other factors like changing refraction (altitude above horizon) and tele-

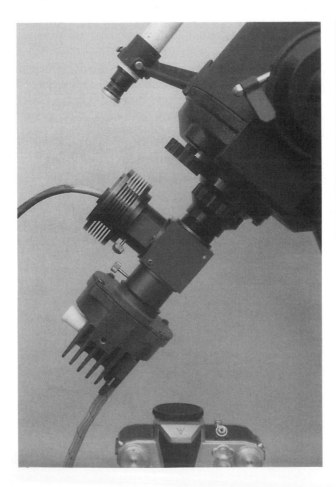

Figure 7.5 The *SIBIG ST-4* CCD autoguider (upper camera) will lock on to a star for the duration of a long exposure, constantly adjusting the telescope's drive motor speed to ensure pin-sharp images on the (lower) CCD or film camera.

scope deformation or 'sag' throughout an exposure that the PEC is unaware of. Electronic guiders can take over where the drive (and PEC) leave off for fully automated photography. The guider shown in Figure 7.5 constantly adjusts the drive motor's speed. Essentially, the telescope can be left unattended (but under a watchful eye) throughout the duration of the exposure.

Early (1980s) electronic guiders used a photodiode or set of diodes to sense any drift of the guide star off target and correct the drive motors both in right ascension and declination accordingly. Generally these units were fairly insensitive, needing a brightish star in the target area to be effective. By the 1990s guiders using the ultra-sensitive CCD or charge-coupled device became affordable to amateurs, and this field has blossomed dramatically. Reference to deep-sky pictures published in astronomical magazines reveal the success these devices have had.

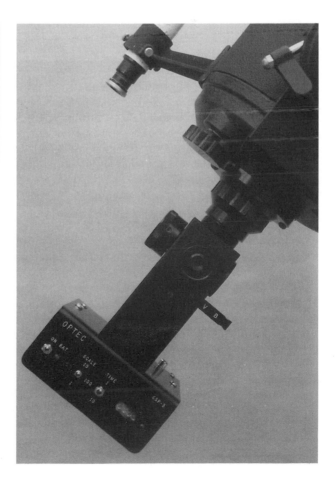

Figure 7.6 The *Optec SSP-3* photometer has a flip mirror for the eyepiece and slider with V and B filters.

Photoelectric Photometry (PEP)

This is an area where the amateur can do science by accurately monitoring the light variations in minor planets or variable stars. This is particularly so on a short time-scale of perhaps minutes or hours where the light variation is of small amplitude. Changes as small as perhaps 0.01 of a magnitude are detectable – better than is ever possible by eye or film.

By using suitable filters (UBV system), stellar colour temperatures can be measured. Much of the equipment has been home-built, but commercial photometers are now available. There are generally two types of detector – those using a photomultiplier tube (PMT) and those using a photodiode, like the Optec model shown in Figure 7.6. The latter is more robust, if currently not as sensitive as the PMT.

CCD Imaging Cameras

The CCD (or charge-coupled device) is a solid-state device consisting of thousands of tiny light-sensitive picture elements or pixels. The CCD imaging camera is placed at the focus of a telescope instead of a film camera (see Figure 7.7). Each pixel converts the incoming photons of light into an electrical charge. At the end of the exposure this charge is transferred through the CCD array, amplified, stored as a picture, and displayed on a computer monitor. Because the image is in digital form it can be stored and manipulated by using the computer as an electronic darkroom.

The CCD has further advantages over film. It is extremely efficient in converting photons to electrons. A 'quantum efficiency' of between 50 and 80% is typical, compared with 2 to 3% for photographic film. The CCD does not suffer reciprocity failure like film, where the effective speed drops markedly under prolonged exposure to the feeble light common in astronomy. The CCD is said therefore to have a 'linear response' where doubling the exposure doubles the data recorded. The CCD's 'dynamic range' or ability to record simultaneously extremely faint and bright part of the picture in a single image is far superior to film. For any given

Figure 7.7 Images from the CCD chip (the rectangle seen centrally within the mounting) are stored digitally on computer disk. A piece of 35-mm film is shown for comparison.

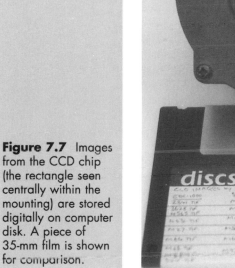

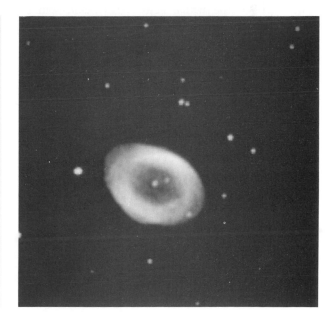

Figure 7.8 The CCD camera can record images many times fainter, much more quickly, than conventional imaging systems. Seen here is M57, the Ring Nebula. Stars to magnitude 19 have been captured in an 80-second exposure.

Picture: Ron Arbour

telescope and exposure, the CCD camera can record stars 100 times fainter (5 magnitudes) than those that can be seen with the human eye, and 16 times fainter (3 magnitudes) than photographic film (see, for example, Figure 7.8).

These factors have revolutionised both professional and now amateur astronomy as CCD prices tumble. Amateurs can now investigate areas previously the sole domain of the professional – the Universe is so much bigger through CCD 'eyes'. Even town dwellers blighted by light pollution can produce results superior to dark sky sites using photography. Admittedly, the field of view imaged on to current amateur CCDs is very small compared with that of film. The active area may vary from a tiny 2.4 mm square (see Figure 7.9) to 13.5 mm square (at present; this is likely to improve with the spin-off from CCD technology, but will need ever-larger computer memories to store the huge picture files). The pixels are also rather coarse, 9 to 27 micro-metres square, compared to the 5 micro-metre resolution of films like Kodak 2415. In practice this is not a problem if the target, CCD camera and telescope are sensibly matched.

CCDs fall into two general categories – the *frame transfer chip* and the *interline transfer chip*. In the frame transfer chip virtually the whole active area is exposed to light – which is potentially ideal. However, at the end of the exposure the image is transferred pixel by pixel down through the whole active array, and will be smeared during readout if it is not shuttered from starlight etc. Full-frame transfer chips have a second but masked area of the chip outside the active area to temporarily store the image before transfer to the computer. In the interline transfer chip this storage area is immediately adjacent to each light-sensitive pixel. The image can thus be transferred sideways instantly without smearing, but of course half the CCD is now masked to light. In practice there is little to choose in terms of sensitivity between the frame and interline transfer chips, and each camera manufacturer and user will have a preference. What is more important is the way the image is sampled during readout to minimise electronic noise that can easily swamp a feeble image. The interline transfer chip is inherently less noisy, and may have a spectral sensitivity more akin to that of the human eye.

Because of the CCD's very high effective 'speed' it is possible to filter the light and still have exposures of modest duration. The CCD's extended sensitivity into

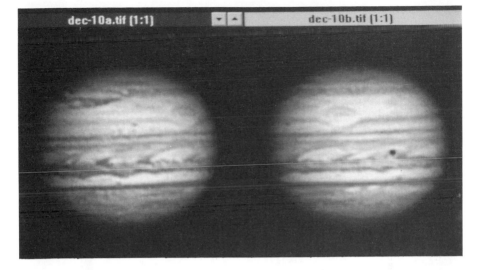

Figure 7.9
Amateur CCD
arrays, at present,
are tiny when
compared with the
imaging area of
35-mm film. This
high-resolution image
of Jupiter was
captured on the tiny
2.4 mm² IC211
chip of the *Lynxx*
CCD camera.

Picture: Dr Don
Parker

the near infra-red could be exploited for an alternative view of the night sky. Potential areas of amateur study include stellar spectroscopy, polarised and narrow-band filtering and full colour planetary and deep-sky imaging through a process called tricolour imaging. The latter area could be very attractive for amateurs. Here, separate exposures are made through red, green and blue filters placed before a monochrome camera. Later, the images are electronically superimposed in the computer and coloured to match the original filters to produce a full colour image on a colour monitor. It needs skills at both the telescope and the computer to get satisfactory results from this technique.

One UK manufacturer (Starlight Xpress) has eliminated many of the problems, with an advanced colour camera which gives high colour saturation results for immediate display after a single exposure. Each block of four pixels over the entire CCD is covered with micro-filters coloured yellow, cyan, magenta and green. From this selection of subtractive filters plus the primary green, all shades and tints can be electronically extracted. Enhancement of the luminance signal (the grey-scale brightness) and chroma signal (colour) can be done separately without cross-contamination. (This is extremely difficult to maintain through the tri-colour technique.) All images are recorded in colour, but are displayed in monochrome unless activated by the computer software. The loss of resolution and sensitivity through the filter mosaic is minimal. The camera thus works in either colour or monochrome. The Starlight

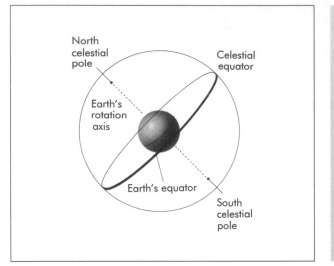

North celestial pole

Celestial equator

Earth's rotation axis

Earth's equator

South celestial pole

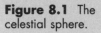

Figure 8.1 The celestial sphere.

poses of doing our calculations. If we do this then the only really serious error occurs with the Moon. It is so near to us that the apparent position of the Moon against the starry background, at any one moment, varies noticeably depending on one's precise position on the surface of the Earth.

The rotation of the Earth on its axis gives the illusion of the celestial sphere turning once a day. As Figure 8.1 should make clear, the points about which the celestial sphere appears to rotate are the projections of the Earth's north and south rotation poles. These points are known as the *north celestial pole* and the *south celestial pole*. Similarly, the projection of the Earth's equator into space is the *celestial equator*.

Great Circles and Meridians

A great circle is any circle drawn over the surface of a sphere, which has the centre of the sphere as its centre. A great circle therefore wraps all around the sphere, and cuts it exactly in half. It is the largest possible circle one can draw on a sphere. The Earth's equator and the celestial equator are both examples of great circles. Of course, any number of great circles can be drawn on a given sphere. Considering the celestial sphere, a great circle that cuts through both celestial poles will intersect the celestial equator at right angles. The equivalent is also true on the Earth. In either case these special great circles are known as *meridians*.

The Celestial Sphere as seen from a Given Location

An observer situated at the Earth's North Pole would have the north celestial pole directly above his head, or at his *zenith*. The celestial equator would skirt his horizon, and the south celestial pole would be directly beneath his feet, through the other side of the Earth, at his *nadir*. If he were situated on the Earth's equator, the celestial equator would arc across the sky, from the *east cardinal point*, through his zenith, to the *west cardinal point*. The north celestial pole would then be at the *north cardinal point* on the horizon and the south celestial pole would be at the *south cardinal point* on the horizon. The cardinal points are the directions of *true* North, South, East and West.

Figure 8.2 illustrates the measurements of *azimuth* and *altitude* as referred to a celestial body. The altitude of a celestial body is the angular distance, *a*, of the body from the observer's local horizon, as measured from the observer. Clearly, an object with an altitude of 90° is at the observer's zenith. The object's azimuth is the angular distance, *A*, measured along the horizon, eastwards from the north cardinal point to the great circle that passes through the zenith and the body. Once again, the measurement is made from the point of view of the observer.

If the observer were located at some particular position on the surface of the Earth, say at a latitude of ⌀°N, then the sky he or she would see would be that illustrat-

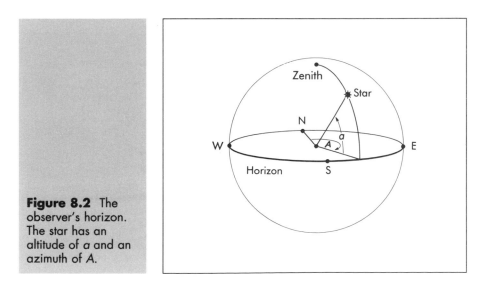

Figure 8.2 The observer's horizon. The star has an altitude of *a* and an azimuth of *A*.

Meridian, right ascension is always measured eastwards from the Vernal Equinox. Also, unlike terrestrial longitudes, right ascensions are seldom measured in degrees. They are usually expressed in units of time. However, before we go further, it is about time to deal with the use of a pocket calculator for making calculations involving angles and trigonometrical functions.

Pushing Buttons and Getting Numbers

I am assuming that you have a fairly cheap and simple pocket calculator that is none the less capable of operating with simple trigonometrical functions (sines, cosines and tangents).

Let us say that you are doing a calculation which requires you to find the sine of the angle 43° 24' 52". How do you do it? The first step is to convert the angle to degrees and decimal fractions of a degree. Remember, there are 60 arc seconds in one arc minute and there are 60 arc minutes in one degree.

First take the number of arc seconds (52) and divide by 60 to find the number of arc minutes this represents. The answer is 0.867. Write this figure down, or store it in the calculator's memory. Next take the number of arc minutes (24) and add on the figure you have calculated. The result is now 24.867. Divide this figure by 60, and you now have the number of degrees this represents. The answer is 0.4144. Finally add on the number of degrees in the original number (43).

Hence 43° 24' 52" = 43.4144°. All that remains is to press the sine button on the calculator. The required answer is 0.6873.

As a second example let us say that, at the end of a calculation, you get a result like cos δ = 0.678 506, and you need to know the value of δ. On most calculators there is an inverse, 'INV', or an arc, 'ARC', button which, after entering the number, you have to press before pressing the cosine button in order to get the angle. Other calculators might have a separate 'COS^{-1}' button. If you try this with your calculator, you should get the answer δ = 47.272 993°.

That answer may well do. If, however, you want to convert it to degrees, arc minutes, and arc seconds you have to do the reverse of the conversion given in the earlier example. Assuming your calculator is still reading 17.272 993, first note down the number of degrees (47) and then subtract it from the reading (giving, of course, 0.272 993). Multiply this figure by 60 to convert it to the number of arc minutes. Your calculator will now read 16.379 58. Write down the whole number of arc minutes (16) and then subtract 16 from the display (giving 0.379 58). Multiply this figure by 60 and you then have the number of arc seconds, 22.7748. Rounding to the nearest whole number, this is 23.

Hence 47.272 993° = 47° 16' 23". More expensive scientific calculators – they are still cheap! – may well have a facility to make this conversion (both ways) automatically, which saves time and effort.

Beware the apparent precision offered by your calculator. Just because its display presents you with an eight-digit answer, that does not mean that your answer is necessarily as accurate as that. In particular, if you have a number like 23.7°, the implied accuracy of the figure is plus or minus 0.05°. Converting it to 23° 42' 00" is then clearly unjustified (even writing it as 23° 42' is questionable, as the accuracy implied in the original figure is only to plus or minus 3').

One final piece of advice I offer is to avoid rounding up or down figures until the very end of your calculation. Otherwise, unnecessary errors may well accumulate, and then be multiplied up to much larger values during the calculation.

Maximum and Minimum Altitudes

The condition for a celestial body to be *circumpolar* (never set below the level horizon) is easy to understand by looking at Figure 8.5. For a celestial body to be circumpolar from a location in the northern hemisphere, its declination must be greater than the co-latitude (90° – ϕ) of the observation site. In general, the maximum possible altitude of the body is given by

maximum altitude = $(90° – \phi) + \delta$.

If the answer comes out to greater than 90° then the star is beyond the zenith at that point. Subtract it from 180° to find the height above the *northern* horizon.

The body is then said to be at *upper culmination*. The

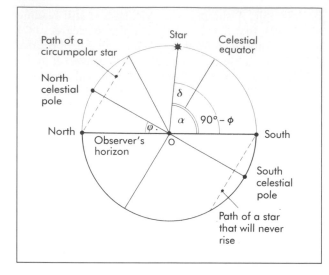

Figure 8.5 The sky from a latitude of $\phi°N$, showing the paths of stars which never rise, and those which never set. The altitude of a star at upper culmination is shown as angle a, which is equal to $(90° - \phi) + \delta$. For a star to be circumpolar it must be no further from the pole than ϕ. Therefore its declination must be less than $90° - \delta$.

minimum altitude of the body (it is then said to be at *lower culmination*) is given by

$$\text{minimum altitude} = \delta - (90° - \phi).$$

If, after putting in the figures (taking care to include the sign of the declination: negative for bodies south of the celestial equator), you get a negative answer then this tells you that the object is below the horizon.

For example, let us work out the maximum altitude of a star of declination $\delta = -73°$, as seen from an observation site at 40°N

$$\text{maximum altitude} = (90° - 40°) + (-73°) = -23°.$$

The result means that, at best, the star is still 23° below the horizon.

Can you see what the formulæ are for the maximum and minimum altitudes of celestial bodies, as seen from southerly observation sites?

Time

Solar and Universal Time

Since the Earth orbits the Sun with a period of one year, the Sun appears to move against the backdrop of fixed stars, going once around the sky every year. The Earth's orbit is slightly elliptical, rather than being perfectly circular, resulting in the Earth–Sun distance and the speed of the Earth in its orbit varying a little over the course of the year. Due to this fact, astronomers have invented a fictitious *Mean Sun*, which is defined to go at a uniform speed around the celestial sphere. Sometimes the real Sun lags a little behind the Mean Sun. At other times it is a little in front.

For an observer in the Earth's northern hemisphere, the Sun reaches upper culmination when it crosses the observer's meridian to his south. The time of crossing defines local noon. In fact, we should be a little more precise. In setting up our system of *solar time*, astronomers have utilised the Mean Sun in order to make the days all of equal length.

Owing to its being an important centre of navigation, astronomers define *Greenwich Mean Time* with reference to the Mean Sun as seen from the Greenwich Meridian (the zero of the Earth's longitude system). The Greenwich Mean Time (GMT) is defined to be 12^h 00^m 00^s exactly at the instant the Mean Sun transits (crosses) the observer's meridian at Greenwich. 24^h 00^m 00^s of solar time then elapses before the next transit. Despite other locations around the world having their own time zones (usually a whole number of hours plus or minus the Greenwich Mean Time), GMT has become sufficiently important to be referred to as *Universal Time* (UT).

Sidereal Time

While a time system based on the apparent movements of the Sun is obviously advantageous for civil use, a time system fixed to the apparent motions of the stars is of great use to astronomers. The *sidereal day*, defined to be 24^h 00^m 00^s of *sidereal time*, is the time interval taken for successive passages of any given star across the observer's meridian. The solar day is slightly longer than the sidereal day because the Earth has to turn exactly once on its axis in order to bring a given star back to the observer's meridian after transit. The Earth, because it is also moving in a curved path around the Sun, has turned a *little more* than once on its axis in order to bring the Sun once again into transit.

In terms of solar time, the sidereal day is 23^h 56^m 04^s long. 00^h 00^m 00^s sidereal time is defined to be the exact moment when the Vernal Equinox transits the observer's meridian at upper culmination.

Of course, an observer at another longitude will not see the Vernal Equinox transiting his meridian at the same time. Therefore, the *local sidereal time* at a given location (LST) is defined to be 00^h 00^m 00^s when the observer, at that location, sees the Vernal Equinox transiting his meridian at upper culmination. Another distinction we can make is the *Greenwich Sidereal Time* (GST), which is defined to be 00^h 00^m 00^s when the Vernal Equinox transits the observer's meridian, at upper culmination, as seen from the Greenwich Meridian.

Since zero hours sidereal time is defined as the moment of transit of the Vernal Equinox across the observer's meridian, the sidereal time at any later moment is simply the sidereal time elapsed since then. If crude accuracy (to a couple of minutes or so) is all that is required, then this can be taken to be equal to the solar time elapsed. Otherwise the slight difference in the lengths of the solar hour and the sidereal hour (9.83 seconds) must be allowed for by calculation, or by consulting conversion tables in an ephemeris.

In any case, the Greenwich Sidereal Time, at any instant, can be found from an ephemeris. To find the local sidereal time, you will then have to include a correction for your longitude, using the following equation

$$LST = GST - \lambda$$

where λ is the longitude, *measured in units of sidereal time*. To make the conversion, 15° of longitude corresponds to 1^h 00^m 00^s. λ has a positive value for longitudes measures westwards from the Greenwich Meridian.

Hour Angles

A celestial body can be considered to lie on its own meridian fixed to the sky. This meridian is known as the *hour circle* of the body. Note that the hour circle is fixed with respect to the stars. Therefore it is moving with respect to the observer.

Figure 8.6 illustrates the measurement of *hour angle*. When the celestial body is on the observer's meridian, as seen from a given location, the *local hour angle* (LHA) of the body is then zero. At any other time the local hour angle of the celestial body is equal to the sidereal time that has elapsed since the celestial body transited the observer's meridian.

As Figure 8.6 shows, the local hour angle of the celestial body can be expressed as a distance (more exactly, as an angle), measured westwards from the observer's

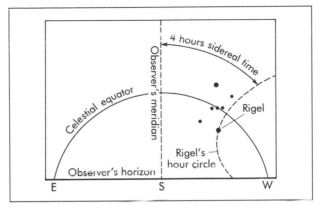

Figure 8.6 The measurement of hour angle. Note that the peculiarly distorted look to the sky is a result of the map projection used. The hour angle of the star Rigel is shown as 4^h 00^m. If the right ascension of Rigel is 5^h12^m, what is the local sidereal time at the instant shown?

meridian. The hour circle of a celestial body will move westwards across the sky by 15° with every hour of sidereal time that passes (the angle being measured along the celestial equator, from the Earth at the centre of the celestial sphere).

Note that the local sidereal time is equal to the local hour angle of the Vernal Equinox. Also, the Greenwich Sidereal Time (GST) is the Greenwich Hour Angle (GHA) of the Vernal Equinox (the hour angle of the Vernal Equinox as seen from the Greenwich Meridian).

Right Ascension

'Right Ascension' may sound like a strange unit, but it has its origins in practical astronomy. If you stand next to a telescope that is supported by an *equatorial* mounting that enables it to track the stars in their apparent motion across the sky (see Chapter 3), and face north, then the right-hand side of the mounting will be ascending – at least in the northern hemisphere, where most early telescopic astronomy was done.

The Vernal Equinox is defined to have a right ascension of zero. Any object on the same hour circle will also have a right ascension of zero. Such an object will transit the observer's meridian (at upper culmination) at 00^h 00^m 00^s LST. Any object on an hour circle that is 15° eastwards of the first (measured along the celestial equator, from the centre of the celestial sphere), will transit the meridian at 1^h 00^m 00^s LST. An object that lies on an hour circle that is 37°.5 east of that which contains the Vernal Equinox will transit the observer's meridian at 2^h 30^m 00^s LST, etc.

The point is that the right ascension of a celestial body can be expressed as the sidereal time which has elapsed between the Vernal Equinox transiting the observer's meridian and the hour circle on which the body lies.

This why it is convenient to measure right ascension in units of sidereal time, rather than in degrees as for declination.

Consider a numerical example: if the LST is $8^h\ 00^m\ 00^s$ at a given instant, a celestial body with this right ascension will be on the observer's meridian (at upper culmination) at that instant. If the celestial body has a right ascension of, say, $6^h\ 00^m\ 00^s$ it will then be $2^h\ 00^m\ 00^s$ west of the meridian, at that same instant (it will have transited two hours earlier). Its hour angle will be equal to $2^h\ 00^m\ 00^s$.

If, instead, the celestial object's right ascension is $10^h\ 00^m\ 00^s$, the object would lie $2^h\ 00^m\ 00^s$ east of the meridian, at that same instant. Since hour angle is measured westwards from the Vernal Equinox, we would properly say that the hour angle of the object is then $22^h\ 00^m\ 00^s$.

In general, the following equations relate hour angle (HA) to right ascension (α) and sidereal time (ST)

$$HA = ST - \alpha ;$$
$$ST = HA + \alpha .$$

Distinguishing between these quantities, as measured from an observer's local meridian and from the Greenwich Meridian, we have:

$$LHA = LST - \alpha ;$$
$$LST = LHA + \alpha ;$$
$$GHA = GST - \alpha ; \text{ and}$$
$$GST = GHA + \alpha .$$

When using these equations, if the answer comes out to be more than $24^h\ 00^m\ 00^s$ then subtract this amount from it. The result is the answer you are looking for.

Remember that the local hour angle of a celestial body will be zero when it is on the observer's meridian and the Greenwich Hour Angle is zero when that object transits the meridian as seen from Greenwich.

Be careful when adding or subtracting times. If numbers have to be carried, remember that if you take one from the 'hours' column, you must add 60 to the 'minutes' column. Similarly, 60 must be added to the 'seconds' column for every one minute taken from the 'minutes' column. Of course, the reverse applies for numbers carried the other way. Obvious? Yes. However, this trivial mistake is only too easy to make when we are all so used to manipulating decimal numbers.

Examples

Use the examples in this chapter to test whether or not you can really do it!

Calculation 1

I am writing these words on 1994 February 16^d. Tonight I wish to observe M48, the open cluster in Hydra. I look up its coordinates in my old copy of *Norton's Star Atlas* and they are: $\alpha = 8^h\ 11.2^m$ and $\delta = -05°\ 38'$. My observing site is situated at a latitude of 50.5°N and a longitude of 0.375°E. I want to know the following:

a Assuming M48 is visible tonight, when will it reach its greatest altitude?
b What will that altitude be?

Answers

a M48 will achieve its greatest altitude when the sidereal time is equal to its right ascension (i.e. the LHA = zero in the equation LST = LHA + α).

Opening up my copy of the BAA *Handbook* I find that at $00^h\ 00.0^m$ UT on February 17^d (midnight, $24^h\ 00.0^m$ on February 16^d) the Greenwich Sidereal Time is equal to $9^h\ 47.0^m$.

Using   LST = GST − λ,
then   LST = $09^h\ 47.0^m$ − (-1.5^m).

Thus, my longitude, 0.375°E, is equivalent to -1.5^m of sidereal time, i.e.

LST = $09^h\ 48.5^m$.

Already I can get a rough idea of where the object will be. At midnight tonight it will be approximately 1^1/$_2$ hours west of my local meridian, having achieved upper culmination, when it transited the meridian, about 1^1/$_2$ hours before.

Let us now work out the time of meridian crossing a little more precisely. At $00^h\ 00.0^m$ UT:

LHA = LST − α,
LHA = $09^h\ 48.5^m$ − $8^h\ 11.2^m$,

i.e.

LHA = $01^h\ 37.2^m$.

Hence M48 transits the meridian, at upper culmination, at 01ʰ 37.2ᵐ (measured in sidereal time) before midnight tonight. If I want to be ultra-precise then I can convert this time interval into solar time. To do so I have to subtract 0.3ᵐ, giving 01ʰ 36.9ᵐ (from a convenient conversion table in my ephemeris, although the calculation is easy to make).

Hence:

Universal Time of the upper culmination of M48 = 22h 23.1m UT (24h 00.0m – 01h 36.9m).

b The maximum altitude of M48 above the horizon is given by:

Maximum altitude = $(90.0° – \phi) + \delta$,
$= (90.0° – 50.5°) + (–05.6°)$

(–05° 38' converts to –05.6°, to one decimal place).

Thus:

Maximum altitude of M48 = 33.9°.

Interconverting Altitudes, Azimuths, Declinations, and Hour Angles

As we have seen, the maximum and minimum altitudes of a celestial body can be easily calculated at the times when the body transits the meridian at upper and lower culminations, respectively. The azimuth is then 180° at upper culmination and 0° at lower culmination, as viewed from the Earth's northern hemisphere (the reverse is the case for a southern observation site). What is the altitude, a, and azimuth, A, of a given celestial body at other times? One also might like to find a celestial body's declination, δ, and hour angle, h (from which the body's right ascension can be derived), from its observed altitude and azimuth. The latitude of the observation site is ϕ. The following useful formulæ can be derived using spherical trigonometry:

$\sin a = \sin \delta \times \sin \phi + \cos h \times \cos \delta \times \cos \phi$;
$\cos a = (\sin \delta \times \cos \phi – \cos h \times \cos \delta \times \sin \phi) / \cos A$;
$\cos a = (–\cos \delta \times \sin h) / \sin A$;
$\sin A = (–\cos \delta \times \sin h) / \cos a$;
$\cos A = (\sin \delta \times \cos \phi – \cos h \times \cos \delta \times \sin \phi) / \cos a$;
$\sin \delta = \sin a \times \sin \phi + \cos a \times \cos A \times \cos \phi$;

$\cos \delta = (\sin a \times \cos \phi - \cos a \times \cos A \times \sin \phi) / \cos h;$
$\cos \delta = (-\cos a \times \sin A) / \sin h),$
$\sin h = (-\cos a \times \sin A) / \cos \delta);$
$\cos h = (\sin a \times \cos \phi - \cos a \times \cos A \times \sin \phi) / \cos \delta.$

All the quantities in these equations are measured in degrees. To convert hour angles to degrees multiply the number of hours (and decimal fractions of an hour) by 15. Obviously, choose the equation which provides you with the answer you want, given the data you know.

Calculation 2

Continuing from Calculation 1, I expect to be too busy doing other things to observe M48 when it is at its highest. However, I do expect to be free at midnight (same date and observation site as for question 1). To what altitude and azimuth should I set my 'quick-look' Dobsonian telescope, in order to find M48 at this later time?

Answer

The first step is to calculate the local hour angle of M48 at midnight. This was already done for question 1. The hour angle is $1^h 37.3^m$.

Let us first list the quantities we know:

$\phi = 50.5°,$
$h = 01^h 37.2^m = 01.620\ 000^h \times 15 = 24.30°,$
$\delta = -05° 38'.$

Then, using $\sin a = \sin \delta \times \sin \phi = \cos h \times \cos \delta \times \cos \phi,$

$$\sin a = (-0.098\ 16)(0.771\ 62)$$
$$+ (0.911\ 40)(0.995\ 17)(0.636\ 08)$$
$$= 0.501\ 18$$
$$\therefore \qquad a = 30° 05'.$$

And, using:

$\sin A = (-\cos \delta \times \sin h) / \cos a,$
$\sin A = (-(0.995\ 17)(0.411\ 51)) / (0.865\ 30)$
$\sin A = -0.473\ 27;$
$\therefore \qquad A = 208° 15'.$

Hence:

To observe M48 tonight at midnight I must set my Dobsonian telescope to an altitude of 30° 05' and an azimuth of 208° 15'.

Precession and Nutation

So far, I have tacitly ignored one 'fiddly bit' in dealing with calculating celestial coordinates. As the Earth spins in its diurnal motion its axis of rotation (inclined at an angle of $23^1/_2°$ to the perpendicular to its orbital plane) does not remain perfectly fixed in direction. Instead, it *precesses* in the same manner that a child's spinning top does. The axis, in fact, sweeps round once every 25 800 years. Consequently the positions of the celestial poles do not remain fixed with respect to the stars, but sweep through circles of diameter 47° in this time. For instance, in about twelve thousand years time Vega will then be the North Pole star!

Since the positions of the celestial poles are 'anchor points' for the celestial coordinate system, it follows that any celestial body's right ascension and declination will vary from year to year as precession takes place. Hence the coordinates of a celestial body are given for a particular *epoch*. My old edition of *Norton's Star Atlas* gives all its positions in '1950.0 coordinates'. The most modern edition gives '2000.0 coordinates', but is less convenient to use.

Most observers, who simply want to set their telescopes on a particular target, need not worry too much about precession. Just using 1950.0 or 2000.0 coordinates will be good enough. However, if greater precision is required (perhaps for locating a faint variable star or tracking down an elusive asteroid) then the following formulæ can be used:

Annual precession in right ascension, $\Delta\alpha$:

$$\Delta\alpha = 3.0730^s + 1.3362^s \times \sin \alpha \times \tan \delta.$$

Annual precession in declination, $\Delta\delta$:

$$\Delta\delta = 20.043'' \times \cos \alpha.$$

In the foregoing equations α and δ are the tabulated values of right ascension and declination. The corrections to these values, $\Delta\alpha$ and $\Delta\delta$ (found from the equations), are each multiplied by the number of years forward from the tabulated date (for instance, 47.5 if the coordinates are tabulated for 1950.0 and you want the values at 1997.5) and the corrections are then simply added to the tabulated values. Remember to subtract the corrections if you want to precess backwards.

A much smaller correction is necessary for *nutation*. The Earth's axis, as well as precessing , also 'wobbles' slightly as it does so. However, the amplitude of nutation is very small (amounting to about 10 arc seconds in total – it is made up of three separate components) and so can be ignored by all except those working in precision astrometry.

I have, of course, been able to give you only the basics of astronomical calculations in this short chapter. It is important to understand the fundamentals, but I don't want to give you the idea that amateur astronomy – at more than a basic level – is difficult and mathematical, for it isn't. It is nevertheless important to understand the fundamental concepts. The calculations themselves are not hard to do, and if you are the owner of a personal computer (if you are interested in astronomy, the odds are that you will be) then there are some excellent programs available that will do all the work for you, and display the results as a table, a print-out, an on-screen picture, or even a printed star chart.

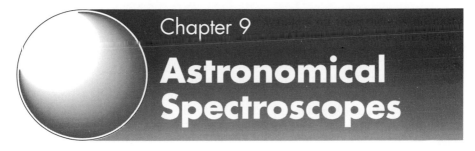

Chapter 9

Astronomical Spectroscopes

C. R. Kitchin

Spectroscopy is the 'neglected art' as far as most amateurs astronomers are concerned. Amongst professional astronomers, however, it is, by contrast, the single most useful and used technique available. At a minimum estimate, 75% of all our knowledge in astrophysics derives from observing spectra.

Surprising as this neglect of such a fundamental technique may seem, it is not difficult to find reasons for it: traditional spectroscopy requires elaborate and heavy equipment to be attached to the telescope; exposures, even for bright stars, can be very long; it needs expensive ancillary equipment and access to obscure data to analyse the resulting images; the theory of spectroscopy appears to be complex and difficult; and, last but by no means least, there is little or no inexpensive equipment which is commercially available for small telescopes, so the observer must design and construct his or her own.

Such a catalogue of woes goes a long way to explaining the unpopularity of spectroscopy among amateur astronomers. Nevertheless, there are several major 'pluses' for the technique. First, the widespread availability nowadays of CCD detectors with a hundred times the sensitivity of photographic emulsion reduces exposure times to reasonable levels (another way of looking at this is that an 8-inch telescope used with a CCD gives a light grasp equivalent to that of an 80-inch telescope of only a few years ago!). Second, guiding is very easy. The star simply needs to be kept on the entrance slit to the spectroscope; if it wanders off then there is no nasty trailed image – you just bring it back on to the slit and continue the exposure. Most importantly, the potential information in a spectroscopic image is orders of magnitude greater than that in a direct photograph, so even if it has taken a long time to acquire, it is more than worth the effort. Finally, the theory of spectroscopy does not have to be understood in its entirety, before you may do useful work, and in any case it is actually no more difficult than astronomy often appears to an outsider.

I hope that this chapter will show that you can do spectroscopy with small telescopes, and will convince some readers to try out the technique for themselves.

Spectroscopes

There are many ways of producing a spectrum: prisms, diffraction gratings, Fabry–Pérot etalons, echelle gratings, Fourier transform spectroscopes, etc.; but only the first two are currently likely to be of interest for use on small telescopes. The prism splits the light up into its component wavelengths because the material from which it is formed has different refractive indices for different wavelengths. The diffraction grating splits up the light through interference effects. The details of the theory of both devices may be found in many standard texts on physics and optics (see the Bibliography at the end of this chapter). Here, we concentrate on their practical application for use in astronomy.

The layout of the basic spectroscope, using either a prism or a diffraction grating, is shown in Figure 9.1. The entrance slit is required in order to produce a spectrum of reasonable purity and to reduce background contamination. The final spectrum is made up of overlapping monochromatic images of the slit – the narrower the slit, therefore, the more pure the spectrum. It is because of the use of a slit that the light or dark portions of the spectrum appear as lines. For some purposes, the slit may be dispensed with (see below), giving a much more efficient system. The spectral features then appear as monochromatic images of the original object. The collimator provides a parallel beam of light to the prism or grating, while the imaging element focuses the spectrum onto the detector. Most spectroscopes are variations upon this basic theme. For example, several prisms may be used to increase dispersion, the lenses may be replaced by mirrors, or the imaging element by

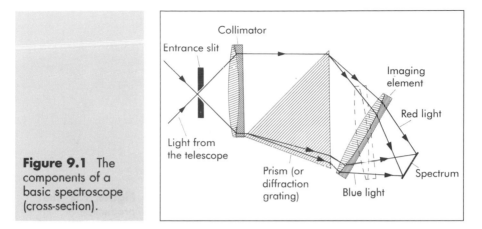

Figure 9.1 The components of a basic spectroscope (cross-section).

a Schmidt camera, etc. The use of a curved reflection grating which can eliminate most of the other components is, however, beyond the scope of this chapter,

Three parameters govern the performance of a spectroscope: spectral resolution, dispersion, and throughput. The spectral resolution is defined as:

$$R = \lambda/\Delta\lambda,$$

where λ is the wavelength and $\Delta\lambda$ is the separation of two wavelengths which may just be distinguished from each other.

For a 60° prism a few centimetres in size, made from dense flint glass, R will have a value between 5000 and 10 000. In the visible, at 500 nm, two wavelengths somewhat less than 0.1 nm apart should therefore be separable. For a diffraction grating, the resolution is given by

$$R = Nm,$$

where N is the total number of lines in the grating and m is the spectral order (see below).

With a modest grating 20 mm across and with 500 lines per millimetre, used in the first order, the spectral resolution is thus 10 000, which is comparable to that of the prism.

If the spectral resolution is to be fully realised, the spectrum must be spread out (or dispersed) sufficiently for the detector to distinguish two just-resolved wavelengths. For a prism made from dense flint glass the dispersion is given near 500 mn by

$$(\Delta\lambda / \Delta x) - (10 / f),$$

where f is the focal length in metres of the imaging element and Δx is the distance in mm along the spectrum

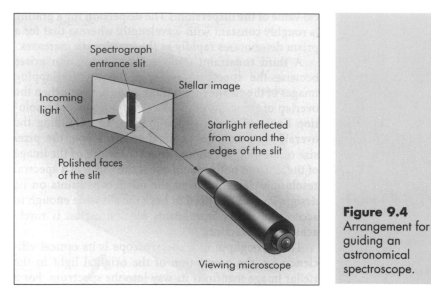

Figure 9.4
Arrangement for guiding an astronomical spectroscope.

direct most of the light falling upon them into the desired spectrum.

For use as an astronomical spectroscope, the basic spectroscope requires a system to allow guiding on the star. This is most usually accomplished by viewing the slit. The outer faces of the slit are polished optically flat, and viewed through a low-power microscope. The stellar image will normally be larger than the slit width, and so the overspill may be seen and used for guiding (Figure 9.4). The guiding is complicated by the need to widen the spectrum. If the star's image is kept at a single point on the slit, then the resulting spectrum will be

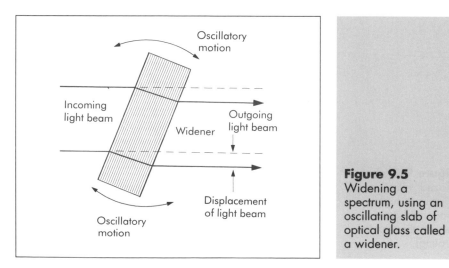

Figure 9.5
Widening a spectrum, using an oscillating slab of optical glass called a widener.

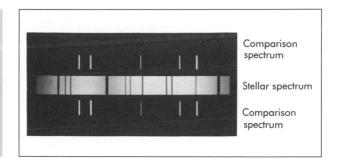

Figure 9.6 A stellar spectrum, set against a comparison spectrum the emission lines of which are of known wavelength.

so narrow that it will be of little or no use. For a simple spectroscope, the easiest way of widening the spectrum is to trail the star's image along the slit. If the length of the slit is aligned parallel to the diurnal motion of the star (by switching off the drive and rotating the spectroscope until the slit is parallel to the drift of the image) then the telescope drive can be set slightly slow or fast. The star's image will then drift slowly, and when it reaches one end of the slit can be brought back to the other end by use of the RA slow motions. Alternatively a widener can be used. This is the system normally found in larger astronomical spectroscopes. A widener is a thick slab of glass with optically flat and polished parallel faces. It is placed into the light beam at an angle, and then rotated back and forth through a few tens of degrees. The displacement of the image by the block (Figure 9.5) causes the light beam to oscillate back and forth, so widening the spectrum.

For advanced work, such as measuring radial velocities, it is necessary to have a comparison spectrum. This is an emission line spectrum, produced by an emission lamp or an electric arc on or near the telescope, which is managed either side of the star's spectrum (Figure 9.6). The wavelengths of the emission lines are known (or can be looked-up in tables – see the Bibliography at the end of this chapter), and so the observed wavelengths of the lines in the star's spectrum can be determined. The wavelength shift of the observed stellar lines when compared with their rest wavelengths gives the star's radial velocity via the Doppler formula

$$v = 3 \times 10^5((\lambda_o - \lambda_r) / \lambda_r),$$

where v is the radial velocity in km s^{-1}, λ_o is the observed wavelength of a line and λ_r is the rest wavelength of a line.

For some purposes, a slitless spectroscope may be used. The commonest form for such a spectroscope is

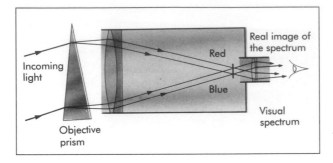

Figure 9.7 An objective prism, the most common form of slitless spectroscope.

the objective prism (Figure 9.7). This is simply a large, low-angle prism placed in front of the objective. The image then contains short spectra for every object in the field of view. Not only does the system have a higher throughput because of the absence of the slit and the other optical components of the basic spectroscope, but many extra spectra can be obtained simultaneously. In the case of a Schmidt camera this could be up to a hundred thousand spectra on a single exposure. The system is thus ideally suited to survey work. The basic objective prism deviates the light beam (Figure 9.7) so that the telescope has to be offset from the object it is viewing. The deviation can be eliminated at the cost of a reduction in dispersion by using two opposed prisms of different glass. This direct-vision objective prism (Figure 9.8) is the inverse of the achromatic lens, since it produces dispersion without deviation.

For the amateur, the objective prism also has the disadvantage of high cost. However a reasonably satisfactory prism may be made by a competent handyman using plate glass for the faces, filled with water (Figure 9.9). If better optical performance is desired then two optically flat windows may usually be bought for less than the cost of an all-glass prism. The lower dispersion

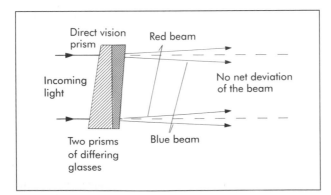

Figure 9.8 A direct-vision prism, used instead of a simple objective prism, eliminates the need for an offset.

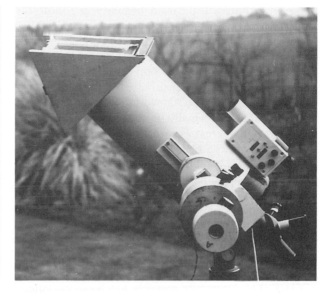

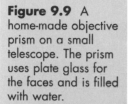

Figure 9.9 A home-made objective prism on a small telescope. The prism uses plate glass for the faces and is filled with water.

Picture © C. Kitchin 1994.

of water compared with glass, however, means that an apex angle of about 45° needs to be used, making the prism heavy for mounting on a small telescope. (The observer would be well advised to incorporate some uncoloured antifreeze into the water within the prism, or risk all the work put into it going for naught when, on a cold night, the water freezes and cracks the glass!)

There are three practicable choices for the detector for a small spectroscope: the eye, photography, or a CCD. It is usually most satisfactory if a permanent record of the spectrum can be obtained. This may then be studied and measured later at leisure (and in warmth and comfort!). However, visual spectroscopy still has its place, particularly for monitoring spectrum variables (see later in this chapter), or just for the pleasure of seeing the spectra and identifying the nature of the objects being viewed. The CCD is perhaps the best choice overall for the detector, though these are still comparatively expensive at the time of writing, and are physically quite small. Furthermore, the devices currently available commercially are most sensitive in the red and infrared. They are much less sensitive (though still more sensitive than photographic emulsion) in the blue part of the spectrum where there is much of spectroscopic interest.

For most observers using a spectroscope on a small telescope, photography is therefore likely to be the chosen detector. The long exposures often required for spectroscopy mean that hypering the film will usually be very advantageous. In other respects, photographing a spectrum differs little from ordinary astronomical photography. The observer should note however, that despite the fact that it is a coloured object that is being photographed, colour film should not normally be used. The three emulsions in such film do not overlap completely in their spectral sensitivities, and so an incomplete spectrum will be obtained. A suitable black-and-white film should normally be chosen, sensitive to the wavelengths of interest. For work into the near infra-red a cheap detector can be built based upon a p–i–n photodiode, and this is then used to scan the spectrum out to about a wavelength of about 1 μm.

Spectroscopes for Small Telescopes

The lack of commercially available spectroscopic equipment for small telescopes means that the intending spectroscopist must be prepared to develop and build his or her own equipment. There are three main approaches to adding a spectroscope to a small telescope; a visual direct-vision spectroscope, an objective prism or a slit spectroscope.

The first of these options, the visual direct-vision spectroscope, is likely to be simplest, quickest and cheapest way into spectroscopy. Small direct-vision spectroscopes are sold, fairly cheaply, for use in schools and colleges to enable chemistry students to identify

Figure 9.10 A small direct-vision spectroscope attached to an eyepiece for visual spectroscopy.

Picture © C. Kitchin 1994.

Figure 9.11 A commercially produced, grating-based spectroscope for use on a small telescope.

Picture © C. Kitchin 1994.

elements via the emission lines produced when a sample is heated in a Bunsen burner flame. Such a spectroscope can simply be attached to a normal eyepiece (Figure 9.10). If the spectroscope contains an entrance slit and/or an eyepiece of its own then these should be removed. Spectra of stars can then be seen with at least the Fraunhofer lines detectable. Extended objects will not normally give a good spectrum with this system because of the lack of a slit. The gaseous nebulæ, however, whose spectra consist of isolated emission lines without a background continuum, will be seen as a series of monochromatic images in the light of each of their spectrum lines.

The objective prism has already been referred to (Figures 9.8 and 9.9). Used with a long-focal-length telescope it will give quite high dispersion spectra, and reveal many more than just the Fraunhofer lines. It may be used visually, or the detector (photographic or CCD camera) normally attached to the telescope can be used to obtain permanent records.

The 'Rolls-Royce' of spectroscopes for a small telescope would be a slit spectroscope with the facility to provide a comparison spectrum. This is not beyond the ability of a good handyman to construct, or (for a price) it may be purchased (Figure 9.11). There is not space here to give detailed instructions on the construction of such a spectroscope and, in any case, construction details will vary depending upon the abilities of the craftsman and the equipment available. Normally, however, at least the prism(s) or the diffraction grating will

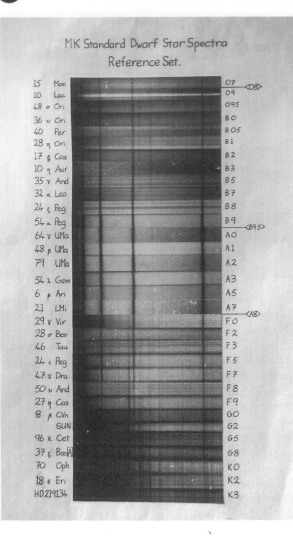

MK Standard Dwarf Star Spectra
Reference Set.

15 Mon	O7 <O8>
10 Lac	O9
48 σ Ori	O9.5
36 υ Ori	B0
40 Per	B0.5
28 η Ori	B1
17 ς Cas	B2
10 η Aur	B3
35 γ And	B5
32 α Leo	B7
24 ς Peg	B8
54 α Peg	B9
64 γ UMa	A0 <B9.5>
48 β UMa	A1
79 UMa	A2
54 λ Gem	A3
6 β Ari	A5
21 LMi	A7
29 γ Vir	F0 <A8>
28 σ Boo	F2
46 Tau	F3
24 ι Peg	F5
47 x Dra	F7
50 υ And	F8
27 η Cas	F9
8 β CVn	G0
SUN	G2
96 κ Cet	G5
37 ς BooA	G8
70 Oph	K0
18 ε Eri	K2
HD219134	K3

Figure 9.12
Stellar spectra
obtained using the
spectroscope shown
in Figure 9.11,
attached to a
16-inch telescope.

Reproduced by
permission of Mrs H
Reeder, University of
Hertfordshire
Observatory.

need to be purchased from an outside supplier. The
design chosen should ensure that the components are
kept rigidly in their correct orientations whatever the
direction of gravitational loading imposed by the
changing position of the telescope. High-quality spectra
(Figure 9.12) may be obtained with such a spectroscope.

The limiting magnitude for direct-vision or objective-
prism spectroscopy will depend upon the throughput,
the dispersion, the widening and the observer's patience,
but a good rule of thumb would be about 3 to 5 magnitudes brighter than the limiting magnitude for equivalent
direct observations. For the slit-based spectroscope, the
limiting magnitude is likely to be from 5 to 10 magnitudes brighter than equivalent direct observations.

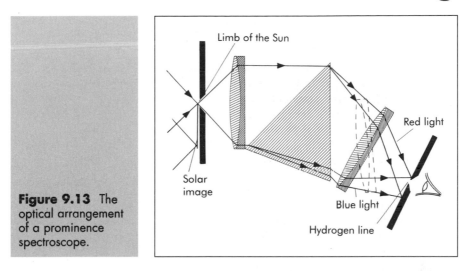

Limb of the Sun

Red light

Solar
image

Blue light

Hydrogen line

Figure 9.13 The optical arrangement of a prominence spectroscope.

For observers interested in the Sun, there is also the prominence spectroscope. This is a slit spectroscope, with a second slit in the focal plane which isolates (usually) the red line of hydrogen known as Hα (Figure 9.13). The entrance slit is set on the limb of the Sun, and prominences can be observed because they emit strongly in the hydrogen line, and so are as bright as the nearby photosphere. Oscillating the slits in phase with each other enables a larger proportion of the solar surface to be covered, and the instrument is then known as a spectrohelioscope.

Spectroscopy on Small Telescopes

At its simplest, spectroscopy on a small telescope can be an extension of touring the universe to admire its diversity. Observations are made visually with a direct-vision or an objective-prism spectroscope. With such a spectroscope, it is the stars, often bypassed on such a tour in favour of nebulæ and galaxies, which become the major players. Their temperature differences can be judged from the different lines and from where the continuum background of the spectrum reaches its greatest intensity. Gaseous nebulæ and comets can instantly be distinguished from distant galaxies because of the former's emission line spectra.

At a slightly more advanced level, spectral classification (Figure 9.12) can be attempted. This would normally be done from photographs or CCD images. However, the early spectroscopists managed without such aids, and visual classification can still be attempted successfully today after some practice. The observer will need to become familiar with the appearance of the spectra of spectral type standards, and either have a very good memory or make good sketches to work against. Atlases of spectral types (see the Bibliography) may help, but spectra are best judged against standard stars viewed through the same equipment.

Some stars, most notably the Cepheid variables, change their spectral type with time. In a Cepheid this may be by as much as an entire spectral class (from F to G and back). A change of spectral type of such a magnitude is easily detectable visually, and could be correlated with the variable's light curve by measurements or estimates of its brightness. For northern hemisphere observers, one of the best Cepheids is η Aql (visual magnitude range 4.1 to 5.3, spectral type range from F6 to G2), while in the southern hemisphere, there is κ Pav (visual magnitude range 3.9 to 4.8, spectral type range from F5 to G5). Polaris (α UMi) is also a Cepheid but with a very small variation in brightness and spectral type. Other spectrum variables may be monitored for activity. Suitable candidates would include novae and supernovae, eclipsing binaries, and emission line stars such as γ Cas and ρ Cyg. Other emission line stars, such as the Wolf–Rayet stars, are worth a look, though the brightest of these, such as $γ^2$ Vel (RA$_{2000}$ 8h 9.6m, dec$_{2000}$ – 47° 21', m_v = 1.7), are only visible to those observers in the southern hemisphere. In the northern hemisphere, the brightest Wolf–Rayet star is HD193793 (RA$_{2000}$ 20h 20.5m, dec$_{2000}$ + 43° 41', m_v = 7.2).

If the apparent magnitude of a star may be measured as well as its spectral type, then your own Hertzsprung–Russell diagram may be constructed. If this is calibrated using the known absolute magnitudes of some of its stars, then it may be used to estimate the distances of stars by the method of spectroscopic parallax. In this method the spectral type is used to determine the absolute magnitude of the star from the H/R diagram, and then together with the apparent magnitude, the distance in parsecs may be calculated, using

$$D = 10^{0.2(m - M + 5)}.$$

For visual spectroscopy, it will be necessary to assume all the stars to be on the main sequence, but with more precise work the luminosity class may also be determinable.

Solar spectroscopy may easily be undertaken with visual spectroscopy. (N.B. Take all the usual precautions required for solar observing.) Many more than just the Fraunhofer lines will be detectable. The observer should attempt to look for differences between the photospheric and sunspot spectra and/or monitor large groups for flare activity which will produce emission components to the hydrogen and other lines. Planetary spectra are, of course, reflected solar spectra, but modified by the properties of the planet's visible surface. A comparison between planets, or between a planet and the Earth's moon may easily be effected when they are near to conjunction with each other. Similarly, the satellites of the major planets may be observed (try comparing Io with Ganymede or Callisto).

If the spectra may be recorded, whether photographically or by using a CCD, then all of the above observations are greatly facilitated. Furthermore, the longer exposures will extend to the range of observable objects to much fainter magnitudes. A permanent record of previous observation helps greatly when looking for small changes in a spectrum. With care, it may be possible to obtain the luminosity classes of stars as well as their spectral classes, thus considerably improving the distances obtained by spectroscopic parallax. It may even be possible to observe the brightest of the quasars (3C273; RA_{2000} 12h 29.1m, $Dec_{2000} = 2° 3'$, $m_v = 12.8$) and see the Hβ line, for example, shifted from the blue–green into the yellow part of the spectrum. By extending the observations into the near ultra-violet, and observing objects (including the Sun) close to the horizon, the absorption due to the Earth's atmosphere is about 75%. About 20% of this absorption is due to ozone. Monitoring changes in absorption can thus enable a watch on the thickness of the ozone layer to be maintained (though this is complicated by the variable contribution to the absorption by dust, especially after volcanic eruptions).

If a slit-based spectrograph with a comparison spectrum is available then many further observations are possible. Most especially, the radial velocities of objects may be measured. This would open up the many thousands of binary stars to detailed study. In some cases, such as the eclipsing binaries, the observations when combined with photometric studies, may be analysed to provide such fundamental data as the star's masses, sizes and orbital radii.

Bibliography

This is a list of some selected books which will enable you to pursue an interest in astronomical spectroscopy further than by simply using this chapter. It is not intended to be an exhaustive list of books on the topic.

Hearnshaw J, *The Analysis of Starlight*. Cambridge University Press (1990). *A history of spectroscopy, useful as a guide to what can be done with small telescopes and simple instruments.*

Jaschek C, *The Classification of Stars*. Cambridge University Press (1987). *A comprehensive account of spectral and other stellar classification systems.*

Kaler J, *Stars and their Spectra*. Cambridge University Press (1989). *An account of astronomical spectroscopy concentrating on spectral types.*

Kitchen C, *Astronomical Spectroscopy*. Adam Hilger, Bristol (1995). *A comprehensive account of both the general theory of spectroscopy and its application to astronomy.*

Kitchen C, *Astrophysical Techniques*. Adam Hilger, Bristol (1991). *A comprehensive account of all types of astronomical techniques and instrumentation, including spectroscopes, and a listing of spectral type standard stars.*

Longhurst R, *Geometrical and Physical Optics*. Longmans, London (1984). *Includes in-depth treatments of the theories of prisms and diffraction gratings.*

Moore C, *A Multiplet Table of Astrophysical Interest*. NBS Technical Note no. 36 (1959). *A very useful list of spectrum line wavelengths and their identification with elements.*

Yamashita Y, Nariai K and Norimoto Y, *Atlas of Representative Stellar Spectra*. University of Tokyo Press, Tokyo (1977). *A comprehensive guide to spectral classification, including photographs of all types of spectra.*

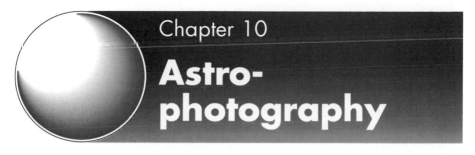

Chapter 10

Astro-photography

Robin Pearce

Whenever I am asked on advice on choosing an astronomical telescope, whether the budget is £200 or £2000, the question always follows, 'Can I take photographs through it?' This is not surprising when one sees the impressive photographs, so often featured in astronomical journals, of galaxies, star-fields and colourful nebulæ.

It has to be said, however, that astrophotography makes the highest demands possible of an astronomical telescope. No other activity requires such a high specification in an instrument, not even CCD imaging. Nevertheless, with a little ingenuity and patience, very pleasing photographs may be taken with quite modest equipment, and it is my intention to encourage readers to achieve this. If advanced astrophotography is undertaken at a later stage then the experience gained will stand you in good stead.

Choice of Camera

Although a wide variety of cameras are available, generally, the modern electronic automatic versions are not suitable for astrophotography, as too many of their functions rely on internal battery power, and long exposure times would not be possible. Often, the automatic features cannot be switched off or over-ridden. And cameras with non-removable lenses are unsuitable, as they cannot be attached to the various accessories that are used with telescopes.

be returned uncut, as the frame divisions will be difficult to judge. 100 ISO is a fine-grained film, and is therefore an excellent choice for lunar and planetary work.
This film also suffers little reciprocity failure; consequently it is an ideal candidate for long-exposure, deep-
sky photography, albeit it has a slight cyan cast (which
is certainly not too objectionable – and if a print or copy
is to be made, the colour balance can be adjusted at
that stage).

Figure 10.2 M57:
Ring Nebula in Lyra.

30 mins at f/6.3,
Fujichrome 100D.
(Photo by the author.)

Fujichrome Provia (400 ISO)

This is an alternative choice, giving similar results to the
100 ISO film, although with an increase in grain at half
the exposure time. It would be a preferable choice for use
with instruments having high focal ratios, such as most
refractors, which usually operate at f/8 or more. (The
relationship between exposure times and focal ratios
will be discussed later.) Fujichrome films do not have a
strong response to the red end of the spectrum, but can
give good results where light skies are encountered.

Kodak Ektachrome 100 and 200

Both of these products are colour slide film, and are

suitable for lunar and planetary work. They are less suitable for deep-sky photography, as they suffer mild reciprocity failure, and are prone to shift towards magenta in long exposures.

At the time of writing (1995), Kodak is replacing the Ektachrome range with the Panther range of slide films, which combine fine grain with high speed. Although these are still an unknown quantity, both the ISO 400 and ISO 1600 films should certainly be investigated.

Kodak Technical-Pan

This black-and-white negative film is the perfect choice for astrophotography; it has no equal in this field. It is exceedingly fine-grained, is capable of very high contrast, and is sensitive in the extreme red of the spectrum, allowing emission nebulae to record well, particularly if a deep red filter is employed. The combination of Technical-Pan and a red filter will permit photography of emission objects even where sky glow is quite severe. Unfortunately, the slow speed (ISO 25, although this varies with different types of development) renders the film too slow for deep-sky photography, where long exposures will be necessary.

The solution to this problem is to have the film gas-hypersensitised. This entails baking the emulsion in a mixture of nitrogen and hydrogen for several days, and then storing it in a freezer. This technique is impractical to undertake at home, but is a service that is available from some astronomical equipment suppliers, who will advise on the film's use and storage.

Do not be deterred from using this film – it will give excellent results in all fields of astrophotography, although you will need to seek out a processing lab to develop and print the film for you.

Processing your own black-and-white film has advantages and is a simple process, requiring only basic equipment and a modest outlay. If you decide to produce your own prints then an enlarger will have to be purchased, but such items can be bought secondhand, and even new models are not expensive.

Kodak T-Max 100 and 400

These two relatively new black-and-white films feature fine grain and give very little reciprocity failure. These

films would be good substitutes for gas-hypersensitised Technical-Pan, although, as they have no response in the extreme red, they would not be suitable for photography of emission nebulae. They would, however, be suitable for photography of galaxies. Processing arrangements would be the same as for Technical-Pan.

Figure 10.3 M31: Andromeda Galaxy.

15 mins at f/4, 300-mm lens, Fujichrome 100D. (Photo by the author.)

Colour Print Films

Few films in this category are suitable for astrophotography. Unless you have a comprehensively equipped darkroom, acquiring satisfactory results is difficult, as two negatives are usually stacked to boost the contrast for the final print. However, two films that are worthy of mention are Kodak Ektar 1000 and Fujicolour Super G 800.

For fast films, these are not too coarse-grained, and will give reasonable results. Ektar 1000 is a good choice of film where light pollution is encountered, because the colour balance shifts towards blue in long exposures; so much so, that it works well with a broad-band filter to producing a natural-looking print. Super G 800 is finer grained than Ektar 1000, and has excellent response to both red and blue. Unfortunately this emulsion is soon to be replaced by Super G Plus, which may not have the same characteristics.

There are, of course, other films that may be used for astrophotography, and manufacturers are introducing new products all the time. Astronomical journals are a good source of up-to-date information, featuring many astrophotos from around the world, usually accompanied by information regarding equipment used and exposure times.

Exposures

Lunar and planetary photographs require short exposures, from about 1/250 of a second for the Moon, and from 1 second for the planets. A certain amount of experimentation will be necessary because, although tables have been published, they are now inaccurate, as film emulsions have changed over the years.

For deep-sky photography there is a simple, but very important, calculation that will give you a 'base' figure for your exposure time in minutes. This calculation is FR^2, where FR is the focal ratio ('f-stop') of the optical system being employed. Whether it is a camera lens or telescope the same rule applies. In other words, for a camera lens operating at $f/1.8$ a base exposure time would be $1.8 \times 1.8 = 3.24$ minutes (say 3 minutes), and for a telescope at $f/6.3$, the figure would be $6.3 \times 6.3 = 39.69$ minutes (40 minutes).

Film types and sky glow have an effect on exposure times, but the concept remains the same. Having calculated the base figure, this could be halved or doubled etc. to achieve optimum results. This concept is well known to photographers, because the f-stops of a camera lens are calculated in the same manner. An increase of one 'stop' requires twice the exposure.

If filters are used then exposure times must be multiplied by the 'filter factor', which may be between ×3 and ×6. Advice should be available from the filter supplier.

Piggyback Photography

This is probably the most underrated form of astrophotography. Many objects are large enough to capture with a modest telephoto lens, piggybacked on an equatorially mounted telescope. Indeed, wide-field photography is only possible with such a system. The prime requirement is the equatorial mounting, preferably with a

Figure 10.4 M27: Dumbbell Nebula in Vulpecula.

45 mins at f/6.3, Fujichrome 100D. (Photo by the author.)

motor drive. The principle is to attach the camera and lens to the back of the telescope for guiding. If the mount is driven and accurately polar-aligned, guiding may not be necessary, in which case the telescope optical tube will not be necessary.

Suitable lenses for this work are those with focal lengths in the 135-mm to 300-mm range and speeds of *f*/4 or faster. Avoid zoom lenses, because they contain too many elements and are not as sharp as prime telephoto lenses. If cost is not a consideration then 'ED' or 'APO' lenses, such as a 180-mm *f*/2.5 or a 300-mm *f*/2.8, will give outstanding results, and even if filters have to be used exposure times will be quite acceptable.

The telescope mounting will need to be polar-aligned as accurately as possible; that is to say, the polar axis must be pointed at the celestial pole and, if it is not driven, a suitable guide star must be located precisely at the centre of the telescope eyepiece for the duration of the exposure. The best way of achieving this is to defocus the guide star, and allow it to drift to the edge of the field, using the telescope controls to re-centre it as it does so.

With a little perseverance, fine astrophotos can be acquired in this manner. It is worth noting that one of the greatest problems encountered in astrophotography is that of achieving perfect focus. With photography through a camera lens the problem is largely eliminated when the lens is set to infinity, although if filters are used then test exposures will have to be made. Some 'autofocus' lenses may also give problems as they can focus past infinity.

Star Trails, Meteors and Constellations

This is the simplest form of astrophotography. The only equipment that is required is a manual 35-mm camera with standard 50-mm lens, cable release and tripod.

Constellations can be recorded in 30-second exposures on ISO 100 or 200 film. Simply aim the camera at the chosen constellation, with the shutter dial set to 'B', and press the cable release, or lock it, for the duration of the exposure. For star trails, adopt the same procedure, but stop the camera lens down to $f/4$ and lengthen the exposure time to 10 minutes.

To capture meteors, a film with high reciprocity failure is preferred, since such a film will respond slowly to low level light, while recording the brief flash from the meteor. A suitable film would be Ektachrome 400. Simply aim the camera at the radiant of the meteor shower and once the meteor is captured on film, terminate the exposure, in order to minimise sky fog.

Lunar and Planetary Photography

Lunar photography is easily possible, even with the quite modest instruments a beginner might purchase. By mounting the camera and lens on a tripod and aiming the low-power telescope eyepiece into it, satisfying results can be obtained, if a moderately fast film is used. This method of photography is known as the *afocal* method. The supplier of your telescope should be able to advise you as to the suitability of your instrument. High-resolution lunar and planetary photography will be a little more demanding of your equipment. A high-quality equatorial mount with motor drive and a good optical system will be necessary. Suitable optical systems would include 3-inch (75-mm) or 8-inch (200-mm) Schmidt–Cassegrain reflectors. Accessories are

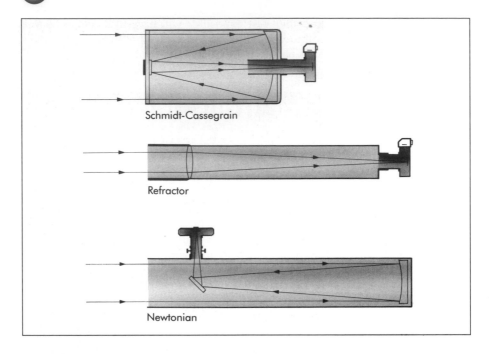

Schmidt-Cassegrain

Refractor

Newtonian

available for all of these telescopes to facilitate this type of photography, where an eyepiece is used to project a magnified image on to the film plane; this is known as positive projection.

Exposures will vary from about 15 seconds down to 1/60th of a second, depending on the brightness of the object to be photographed. Recommended eyepieces for this type of work are Orthoscopics or Plossls, with focal lengths of between 9 mm and 15 mm. Try as many exposure times as you can, and be prepared to use more than one roll of film. When the air is exceptionally steady, and images are better than average, take as many shots as possible, because moments such as these are rare.

Figure 10.5 A Schmidt–Cassegraine reflector, a refractor and a Newtonian reflector. All have a camera body mounted at prime focus.

Deep-Sky Astrophotography

Deep-sky astrophotography through the telescope is technically known as prime-focus photography, or, more accurately, Newtonian-focus and Cassegrain-focus photography when these particular instruments are used. The telescope is coupled directly to the camera body, effectively becoming a large telephoto lens. Figure

Figure 10.6 M42: Orion Nebula.

45 mins at f/6.3, Fujichrome 100D. (Photo by the author.)

10.5 shows a Schmidt–Cassegrain reflector, a refractor, and a Newtonian reflector, all with a camera body at the prime focus.

This is an area of astrophotography where careful consideration must be given to the choice of instrument; there are no short cuts. The one item central to the success of a high-quality deep-sky photograph is the telescope's equatorial mounting. Without this vital piece of equipment, long-exposure photography will not be possible. An equatorial mounting suitable for such work must have the following attributes:

- It should be of good mechanical design, with particular regard to fine adjustments and gears.
- It should be able to accept a high-quality drive system, if this is not included, with fine correction in Right Ascension and declination.
- There should be provision for simple yet accurate alignment of the polar axis, preferably with some

form of visual assistance.
- It should be of sufficient strength for the chosen optical system.

To understand why so much consideration needs to be given to the equatorial mounting, it is helpful to understand what must be achieved.

Objects such as nebulæ, galaxies, clusters and supernova remnants, etc., can be extremely faint. Even the brighter objects such as the Orion nebula and the Andromeda Galaxy only reveal their full beauty when the more elusive detail can be seen. Therefore the intention is to expose the photographic emulsion to the object for as long as is possible, in order to capture the maximum detail.

This can be achieved only if the telescope is aimed accurately at the object for the duration of the exposure, permitting the photons from the image to fall on the film at precisely the same place time after time.

To undertake a lengthy exposure the telescope mounting must first be perfectly polar-aligned. If the telescope is to be permanently sited then this can be achieved without any auxiliary devices, as long as suitable adjustments are available in azimuth and altitude.

The method is time-consuming, but because the mount will not be moved it is a once-only operation. The principle is to aim the polar axis of the mount as near as possible to the celestial pole, which is marked to within 1° by Polaris for the northern hemisphere and σ Octantis for the southern hemisphere. A high-power eyepiece equipped with cross-hairs is then employed, to observe the drift in declination of a star on the meridian, making adjustments in azimuth, and then a star to the east, with adjustments in altitude, until all such drift is eliminated.

A high-quality drive with a hand controller, featuring pushbutton control of both axes, is mandatory, yet even with the greatest precision possible in drive systems the telescope cannot be left unattended during the exposure. Any drive system will possess a characteristic known as periodic error. In order to eliminate this, an auxiliary optical system is employed, to monitor the position of a guide star on the cross-hairs of a guiding eyepiece throughout the exposure, the operator making the necessary corrections with the hand controller. The system employed could be another telescope mounted to the main telescope, or a device known as an off-axis guider, where a prism is inserted into the edge of the light path, diverting the image of the guide star to the cross-hair eyepiece. The advantage of the latter system is that the same optical system used for guiding is used for taking the photograph; this makes it much more accurate, because it eliminates a problem known as differential flexure, which can occur when two separate optical systems are employed.

The method of guiding preferred will, to some extent, depend on the choice of main optical system. Most readers will choose a portable instrument, to take advantage of different observing sites and to fit in with your domestic environment.

Much has been written on the subject of building your own equipment (see, for example, Chapter 5). If you relish an engineering challenge, and have the facilities, by all means proceed.

The modern equatorial mounts that are commercially available from established manufacturers cannot, in my view, be surpassed. They are highly sophisticated, both in engineering and electronic terms, and have some form of polar-alignment assistance together with a versatility that astrophotographers a few years ago could only have dreamed of. If your budget is limited purchase the mount separately, secondhand if at all possible, and build up your system as funds permit.

Choice of Optical Tube

Most importantly, a range of accessories to allow attachment of a camera body must be available for the telescope of your choice. Although this is rarely a problem with commercial instruments, do avoid those that only accept 1.25-inch push-fit adapters that fit the standard eyepiece holder, as the resultant light cone is insufficient to illuminate fully a 35-mm frame.

The next consideration is the focal ratio of the optical system, remembering that exposure times are directly related to this factor. A telescope with a focal ratio of $f/10$ will require exposures some two and a half times longer than one with a focal ratio of $f/6$.

The main telescope designs under consideration are the refractor, the Newtonian reflector and the Schmidt–Cassegrain catadioptric reflector. These were the designs illustrated in Figure 10.5.

Few refractors are suitable for deep-sky astrophotography. Most refractors have achromatic objectives, corrected for the red and blue ends of the visual spectrum, allowing chromatic aberration to be evident in the violet and ultra-violet, an area where photographic film is very sensitive. Refractors also typically operate at $f/10$ to $f/15$, which is too slow for long exposures.

Fully colour-corrected *apochromatic* refractors, particularly those able to operate at focal ratios of $f/5$ to $f/6$, make excellent astrographs, but are costly instruments, especially in apertures above 5 inches, and become very bulky, requiring large and expensive mountings.

The Newtonian reflector is potentially a good choice of instrument for astrophotography, offering more aperture for your money than any other design. Fast focal ratios are possible, $f/4$ to $f/6$ being practical, although below $f/6$ this results in the problem of coma, where star images begin to resemble 'seagulls' towards the edge of the field. While this may not be too objectionable on the photograph, it does present problems if you choose to use an off-axis guider. Therefore a separate guide telescope may be the only method by which you can effect your guiding. Coma-corrector lens systems are available, but are also expensive.

The major problem with the Newtonian reflector is its lack of the ability to back focus, which will be required to bring the image into focus at the film plane of the camera, 2 inches (150 mm) further out than most eyepieces. The best solution to this problem is to fit a photo-visual focuser. Regrettably, at present, there do not appear to be any commercially available telescopes that are suitable for astrophotography without a certain amount of customisation, other than expensive, purpose-built instruments from companies such as Takahashi. Other products could, of course, appear on the market at any time.

The last telescope design I shall deal with is the catadioptric. I will exclude the Maksutov, since few are manufactured and to all intents and purposes they are not suitable for deep-sky astrophotography. The instruments that are familiar to one and all are the Schmidt–Cassegrain telescopes (SCTs), the most popular apertures being 8 inches (200 mm) and 10 inches (250 mm). Each telescope is usually supplied as a complete, fork-mounted instrument, but optical tube assemblies are available for attachment to German equatorial mounts.

The optical and mechanical design of the Schmidt–Cassegrain lends itself perfectly to astrophotography. Photographic accessories are available for all types of

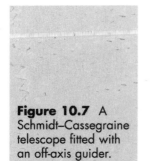

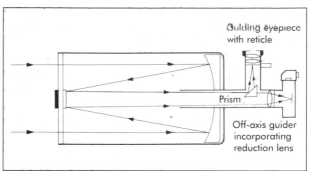

Figure 10.7 A Schmidt–Cassegraine telescope fitted with an off-axis guider.

photography, and the focus mechanism has more than adequate back focus to accommodate them. Guiding is achieved by the off-axis method, as no guide telescope could match the focal length of these instruments (2 metres in the case of the 8-inch (200-mm)).

Although the focal ratio of a typical SCT is $f/10$, reducing lenses are available to yield a focal ratio of $f/6.3$. It is also possible to purchase an $f/6.3$ version, although, in my opinion, this is not as flexible.

The preferred configuration is to use an $f/10$ telescope coupled to an off-axis guider incorporating a reduction lens. The advantage of this system is that the prism used to select the guide star is situated between the telescope and the reduction lens, utilising the full focal length of the telescope, while the camera takes the photograph at the reduced effective focal length, resulting in an increase of 2.5× guiding accuracy. As an example, an 8-inch SCT with a focal length of 2000 mm and a guiding eyepiece with a focal length of 9 mm would produce a magnification of 222×. The camera body, which has a focal length of 1280 mm, would produce a magnification of 25× at the film plane, which is a more than adequate guiding ratio. Figure 10.7 shows a Schmidt–Cassegrain telescope fitted with an off-axis guider which incorporates a reduction lens..

Probably the only disadvantage with the Schmidt–Cassegrain is its cost; it is an expensive optical system to manufacture, although less so than an apochromatic refractor.

One final item that must be purchased is a guiding eyepiece. These usually have a focal length of 9 mm or 12 mm, with a set of cross-hairs illuminated by a red LED. The better versions allow the reticule to be moved within the field of view, enabling easier centring of the guide star.

Techniques for Long-Exposure Astrophotography

A rigid routine is helpful in long-exposure astrophotography – as is a great deal of patience. Having precisely polar-aligned the equatorial mount, ensure that the whole instrument is balanced with all the photographic accessories attached, particularly important with German equatorial mounts.

Having located your chosen subject, compose the view in your camera to your satisfaction, then select a guide star, centring it on the cross-hairs of the guiding eyepiece, with adjustments to whatever guiding system you have chosen. Rotate the guiding eyepiece until the position of the cross-hairs corresponds to the motion of the telescope. Check this by operating one of the controls on the handset while monitoring the guide star. Now you are ready to focus the camera. In order to achieve this, a star must be placed in the camera viewfinder. If no star is available near the subject then I move the telescope in declination until a suitable star appears. Once the camera is focused I operate the opposite declination control and monitor the guiding eyepiece for the reappearance of the guide star. At this point the subject should have returned to the original position.

There are various methods of achieving perfect focus, and I recommend that you try as many as possible. It is widely accepted that the only accurate method is 'knife-edge' focusing. This entails placing a sharp edge against the film rails of the camera (the film plane) and moving it back and forth until the light from the star cuts out all at once. If the effect you see is that of a shadow moving in one direction or another, then you are not at focus. Focus aids are available that utilise a Ronchi grating for the same purpose, the advantage being that the camera need not be opened to perform this task.

A popular, though less predictable, method of focusing is to have the camera equipped with a plain, ground-glass focusing screen and a focus magnifier. Choose a star that is not too bright, in order to minimise light scatter on the focusing screen itself. After checking that all is well, open the camera shutter on the 'B' setting, and commence guiding until the exposure is completed. Needless to say, a record should be made of everything relating to each photograph, in order to assess what progress is being made.

Upon processing the film, examine the results carefully. Anything but pinpoint star images suggests poor focus. Star images that trail indicate a guiding or polar alignment problem. Meticulous attention to setting-up your equipment will pay dividends in the long term.

Many modern telescopes can be interfaced with an electronic autoguider, which utilises a charge-coupled device (CCD) to monitor the guide star, feeding the information to a set of relays which operate the drive motors of the telescope. The benefit here is that longer exposures can be undertaken, without the painstaking task of having to continually monitor the guiding eyepiece.

Solar Photography

One area of astrophotography not covered here is that of solar photography. Any form of solar astronomy must be approached with extreme caution, and I am reluctant to offer advice on the subject in such a short article. Anyone contemplating solar photography will need to be experienced in the use of their telescope, and will have made extensive enquiries through equipment suppliers and reading material before attempting this type of work.

Afterword...

To sum up, the enthusiastic astrophotographer will equip him or herself with whatever is required to achieve their aims, and – I would suggest – a considerable outlay should be anticipated, although the results will more than justify the investment. Astrophotography is highly demanding of both equipment and the astronomer, but with the right instrumentation and attention to fine detail the rewards will be forthcoming.

Bibliography

Arnold H J P, *Night Sky Photography*. George Philip
Wallis B and Provin R, *Manual of Advanced Celestial Photography*. Cambridge University Press

Chapter 11

Astronomical Societies

Patrick Moore

In most walks of life, cooperation is not only desirable, but is also immensely satisfying. This is certainly true in astronomy. Lone observing may be interesting, but in the long run it does not add a great deal to general knowledge, and it is far better to take part in programmes with other observers.

Fortunately there is no problem about this, because there are many astronomical societies – and very few of them impose any limits on age or experience. The British Astronomical Association, founded in 1890, has a record of observational excellence second to none; membership is open to all, and there are various observing and technical sections, each controlled by an experienced director (many of whom have contributed chapters in this book). A bi-monthly journal is issued, as well as circulars giving quick notice of interesting discoveries; most of the sections also issue their own material. Meetings are held monthly, usually in London but also elsewhere.

In England there are also many local societies; a full list is given in the annual *Yearbook of Astronomy*, and frequent reports of their activities come out in the monthly periodicals *Astronomy Now* and *Practical Astronomy*. Many societies have observatories of their own – and the newcomer can usually find someone who will make a telescope available for use. In Northern Ireland there is the Irish Astronomical Society and also Astronomy Ireland. Similar organisations exist in the Channel Islands, the Isle of Man and elsewhere.

Most countries now have national and local societies, and again there are in general no restrictions on membership. Moreover, many amateurs qualify to become members of professional societies; the Royal Astronomical Society, for example, has a number of amateur Fellows (just as there are many professionals who are also members of the mainly amateur BAA).

The overall situation varies in different countries. For instance, the Royal Astronomical Society of Canada and the Royal Astronomical Society of New Zealand have many amateur members, and some of their associated observatories produce work of full professional standard. This also applies to the Astronomical Society of South Africa. Also in Europe, most of the national societies have a large amateur membership. The Société Astronomique de France is a good example.

In the United States the situation is different again. Major organisations, such as the Astronomical Society of the Pacific and the Planetary Society, have amateur members. The Association of Lunar and Planetary Observers is probably the nearest equivalent to the British Astronomical Association. But most large cities and states have local societies of their own and produce work of a very high standard indeed – and *star parties* are much more common than they are in most other countries. To list all the American societies would fill many pages, but regular lists and reports are given in the monthly periodical *Sky & Telescope* – just as lists and reports of British societies are given in the monthly *Practical Astronomy* and *Astronomy Now*.

In every way, membership of an astronomical society is to be recommended without reservation. Only by working with others is it possible to obtain the full benefit, and the full joy, of astronomy as a lifelong hobby.

Bibliography

Books

Kaufmann W, *Universe*. New York: Freeman (1990)
Moore Patrick, *The Amateur Astronomer*. Cambridge University Press (1991)
Ronan C A, *The Natural History of the Universe*. Doubleday (1992)

Periodicals

Astronomy Now (monthly). Pole Star Publications, Tonbridge
Practical Astronomy (monthly). Concept Publications, Milton Keynes

Sky & Telescope (monthly). Sky Publishing Corporation, Cambridge, Mass., USA

The Yearbook of Astronomy (annually). Pan Macmillan, London

Index